Springer Tracts in Modern Physics 83

Ergebnisse der exakten Naturwissenschaften

Springer Tracts in Modern Physics

Volume 66	**Quantum Statistics in Optics and Solid-State Physics** With contributions by R. Graham, F. Haake
Volume 67	**Conformal Algebra in Space-Time and Operator Product Expansion** By S. Ferrara, R. Gatto, A. F. Grillo
Volume 68	**Solid-State Physics** With contributions by D. Bäuerle, J. Behringer, D. Schmid
Volume 69	**Astrophysics** With contributions by G. Börner, J. Stewart, M. Walker
Volume 70	**Quantum Statistical Theories of Spontaneous Emission and their Relation to Other Approaches** By G. S. Agarwal
Volume 71	**Nuclear Physics** With contributions by J. S. Levinger, P. Singer, H. Überall
Volume 72	**Van der Waals Attraction:** Theory of Van der Waals Attraction By D. Langbein
Volume 73	**Excitons at High Density** Edited by H. Haken, S. Nikitine. With contributions by V. S. Bagaev, J. Biellmann, A. Bivas, J. Goll, M. Grosmann, J. B. Grun, H. Haken, E. Hanamura, R. Levy, H. Mahr, S. Nikitine, B. V. Novikov, E. I. Rashba, T. M. Rice, A. A. Rogachev, A. Schenzle, K. L. Shaklee
Volume 74	**Solid-State Physics** With contributions by G. Bauer, G. Borstel, H. J. Falge, A. Otto
Volume 75	**Light Scattering by Phonon-Polaritons** By R. Claus, L. Merten, J. Brandmüller
Volume 76	**Irreversible Properties of Type II Superconductors** By H. Ullmaier
Volume 77	**Surface Physics** With contributions by K. Müller, P. Wißmann
Volume 78	**Solid-State Physics** With contributions by R. Dornhaus, G. Nimtz, W. Richter
Volume 79	**Elementary Particle Physics** With contributions by E. Paul, H. Rollnick, P. Stichel
Volume 80	**Neutron Physics** With contributions by L. Koester, A. Steyerl (This volume includes a Classified Index for Volumes 36-80)
Volume 81	**Point Defects in Metals I:** Introduction to the Theory By N. Breuer, G. Leibfried
Volume 82	**Electronic Structure of Noble Metals, and Polariton-Mediated Light Scattering** With contributions by B. Bendow, B. Lengeler
Volume 83	**Electroproduction at Low Energy and Hadron Form Factors** By E. Amaldi, S. P. Fubini, G. Furlan
Volume 84	**Collective Ion Acceleration** With contributions by C. L. Olson, U. Schumacher
Volume 85	**Solid Surface Physics** With contributions by J. Hölzl, F. K. Schulte, H. Wagner

E. Amaldi S. Fubini G. Furlan

Pion-Electroproduction

Electroproduction at Low Energy and Hadron Form Factors

With 47 Figures

Springer-Verlag Berlin Heidelberg GmbH 1979

Professor Dr. Edoardo Amaldi
Istituto di Fisica dell'Università di Roma, Sezione di Roma dell'INFN,
Piazzale A. Moro 5, I-00185 Roma, Italy

Professor Dr. Sergio Fubini
C.E.R.N., CH-1121 Geneva 23, Switzerland

Professor Dr. Giuseppe Furlan
Istituto di Fisica Teorica dell'Università di Trieste, Sezione di Trieste dell'INFN,
International Center for Theoretical Physics, Strada Costiera 11, I-34014 Trieste, Italy

---

Manuscripts for publication should be addressed to:

Gerhard Höhler
Institut für Theoretische Kernphysik der Universität Karlsruhe
Postfach 6380, D-7500 Karlsruhe 1, Fed. Rep. of Germany

Proofs and all correspondence concerning papers in the process of publication should be addressed to:

Ernst A. Niekisch
Institut für Grenzflächenforschung und Vakuumphysik der Kernforschungsanlage Jülich GmbH
Postfach 1913, D-5170 Jülich 1, Fed. Rep. of Germany

ISBN 978-3-662-15434-2 ISBN 978-3-540-35677-6 (eBook)
DOI 10.1007/978-3-540-35677-6

Library of Congress Cataloging in Publication Data. Amaldi, Edoardo. Pion-Electroproduction — Electroproduction at low energy and hadron form factors. (Springer tracts in modern physics; v. 83) Bibliography: p. Includes index. 1. Hadrons-Scattering. 2. Form factor (Nuclear physics) I. Fubini, S., 1928-. joint author. II. Furlan, G., 1935-. joint author. III. Title. IV. Series. QC1.S797 vol. 83 [QC793.5.H328] 539'.08s [539.7'216] 78-13034

Originally published by Springer-Verlag Berlin Heidelberg New York in 1979.
Softcover reprint of the hardcover 1st edition 1979

2153/3130 — 543210

Preface

During recent years, important progress has been made in understanding the physical properties of hadrons.

The most successful approach is based on the idea that hadrons are composed of more elementary constituents whose intrinsic properties and distributions inside the hadrons determine the observed properties of hadronic matter. Although the existing quark models are still at a rather preliminary qualitative stage, it is likely that they are the first step in the right direction.

Among the important properties of hadrons which do and will constitute a fundamental test of our theoretical models are the static properties of stable hadrons such as, for example, the electromagnetic and weak form factors. This review is, indeed, devoted to a description of the experimental as well as the theoretical problems which are encountered in their determination.

The derivation of proton and neutron electromagnetic form factors from elastic electron scattering on proton and deuteron targets has already been the object of detailed discussions in excellent reviews. We shall thus concentrate our attention on low-energy *inelastic* electron-nucleon scattering and emphasize those form factors which can be obtained from such experiments. In particular we shall deal with the elastic form factor of the pion, the N-Δ transition form factors, and, via current algebra, with the axial structure functions of the nucleon.

Our progress will of course lead us to discuss in some detail dispersion and current algebra techniques which, although less popular today than yesterday, always constitute an invaluable tool for handling a physical situation in which a fundamental dynamical scheme is still lacking.

On the other hand, a discussion of the beautiful but still preliminary attempts which have been devised to compute the form factors from a fundamental scheme lies outside the plan of the present book. The authors express the hope that a general knowledge of the static properties of hadrons might be of much help in future attempts towards a better understanding of hadronic structure.

This book contains material which was available to us before July 1977. For any omissions of important contributions which escaped our knowledge we can only ask for the indulgence of the reader and of those who contributed to the development of the field.

In the course of the writing we have been helped by many friends, and we acknowledge fruitful discussions with M. Beneventano, B. Borgia, A. Donnachie, N. Paver, P. Pistilli, and C. Verzegnassi.

One of us (G.F.) wants to thank Prof. J. Prentki for the warm hospitality at the CERN Theory Division.

Mr. S. Stabile and Mrs. L. Doria-Vici of the University of Trieste and Mr. F. Stazzi and Miss G. Gori of the University of Rome were of great help in technical matters.

Trieste, January 1979 E. Amaldi S. Fubini G. Furlan

Contents

1. Introduction

The term "electroproduction" is currently used for indicating the inelastic collision of a charged lepton ($e^{\pm}$ or $\mu^{\pm}$) against a nucleon (or a nucleus) with production of one or several bosons. For example (single) electroproduction of pions indicates the following reactions:

$$\begin{aligned} e + p &\rightarrow e' + n + \pi^{+} && \text{(a)} \\ &\rightarrow e' + p + \pi^{0} && \text{(b)} \\ e + n &\rightarrow e' + p + \pi^{-} && \text{(c)} \\ &\rightarrow e' + n + \pi^{0} && \text{(d)} \end{aligned} \qquad (1.1)$$

the first two of which - lumped together - were studied for the first time in 1958 by Panofsky and Allton /1/ by observing only the electron scattered inelastically by protons. Similar processes in which, instead of a pion, other mesons are produced have either recently been undertaken experimentally, or will be in the near future. In the present book devoted mainly to pions, we consider also the production from isolated nucleons of pseudoscalar mesons belonging to the same SU(3) octet as the pion. Vector mesons involve a different type of physics and therefore are not considered here. For similar reasons we shall consider only the energy region extending from threshold to the first (πN) resonance included. A few review articles on the same problems, which appeared from 1970 to 1976, are listed in /2-7/.

Processes of type (1.1) are described as due to the exchange of one, two..., any number of (virtual) photons between the lepton and hadron currents. The higher the number of exchanged photons, the smaller is the corresponding contribution expected to be as a consequence of the smallness of the electromagnetic coupling constant

$$\frac{e^2}{4\pi} \equiv \alpha = \frac{1}{137} \quad \text{(in } h = c = 1 \text{ units)}. \qquad (1.2)$$

Thus the two-photon exchange contribution amounts to a few percent correction to the results obtained by considering only one-photon exchange, and the terms of higher order are correspondingly smaller. Therefore the "one-photon exchange approximation" (o.p.e.a.) is usually adopted in most discussions of the subject, particularly in the present review. The corrections due to terms of second and higher order in α could be computed, at least in principle, but at present they are beyond, or at most at the limit of, the present experimental accuracy.

1.1 Properties of the Virtual Photon

In the o.p.e.a. the theoretical description of electroproduction becomes extremely simple and physically enlightening: it can be treated as photoproduction by a virtual photon, whose mass $\sqrt{k^2}$, energy k_0, direction $\hat{k}$, and polarization ε are tagged by the scattered electron.

If we denote by l_1 and l_2 the four-momenta of the incident and scattered lepton (Fig. 1.1), the four-momentum of the single exchanged photon is given by

$$k_\mu = l_{1\mu} - l_{2\mu} \tag{1.3}$$

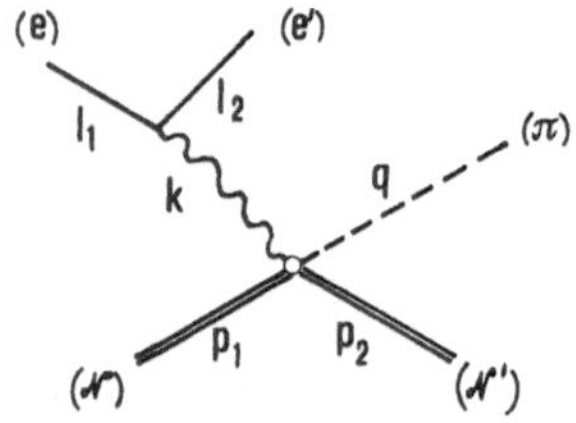

Fig. 1.1. Representation of electroproduction in the one-photon exchange approximation (o.p.e.a.). Notations used for four-momenta are indicated

Two of the three peculiar properties of this photon are clear: 1) it is monoenergetic; 2) it has a mass squared

$$k^2 = k_0^2 - \underline{k}^2 = 2m_l^2 - 2(l_{01}l_{02} - \underline{l}_1 \cdot \underline{l}_2), \tag{1.4}$$

almost always spacelike ($k^2 < 0$). For $\underline{l}_1^2, \underline{l}_2^2 >> m_l^2$, (1.4) becomes

$$k^2 \approx -4l_{01}\, l_{02} \sin^2(\theta_l/2), \tag{1.5}$$

where θ_l is the angle between $\underline{l}_1$ and $\underline{l}_2$.

We see from (1.3) and (1.4) that the values of the energy k_0 and the mass squared k^2 can be chosen at will by adopting appropriate experimental conditions. In particular, by a convenient choice of the kinematical conditions and an appropriate design of the experimental setup, one can determine the desired values of k_o and k^2 and the corresponding accuracies. There is even more. The values $\underline{l}_1$ and $\underline{l}_2$ determine the third - even more important - property of the virtual photon: the virtual photon is polarized with transversal (ε) as well as longitudinal (ε_L) components [see (A.27) and A.28)].

Real monoenergetic photons can be obtained by tagging /8/ or by annihilation of positrons in flight /9/, real polarized photons by bremsstrahlung in amorphous targets observed at an assigned angle /10/ or by means of gamma ray absorption method /11/, or by back scattering of laser beams /13/.

Virtual photons, however, have the advantage that the polarization can be close to 100% and accurately known, and the energy resolution can be made arbitrarily fine. The disadvantage is that the number of virtual photons at one's disposal under reasonable experimental conditions is of the order of 1% of the number of photons for a typical bremsstrahlung beam. Furthermore, the determination of the photon's properties involves the observation, in coincidence with one of the produced particles, of the inelastically scattered electron (tagging). These photons intervene only in electroproduction and a few other related processes involving the inelastic scattering of a lepton or the production of a lepton pair of assigned energies and momenta (Sec. 6.1).

In particular the longitudinal polarization of the virtual photon (1.3) provides a unique tool for the investigation of the structure of hadrons. By a clever choice of the kinematical conditions, one can even isolate the longitudinal photons. For example, only longitudinal photons can produce pions moving in their direction, because they have no helicity to get rid of /14/.

From a purely phenomenological point of view, electroproduction should be seen as a wide class of processes lying between electron-nucleon elastic scattering and photoproduction. This remark is clarified by the following kinematical consideration.

The invariant energy W of the two hadrons present in the final state of any one of the reactions (1.1) can be expressed in terms of k^2 and k_o in the l.f. [see (A.13)]. This expression can be put in the form

$$-k^2 = m_1{}^2 - W^2 + 2m_1k_o, \tag{1.6}$$

which, for a given value of W^2, is represented by a straight line in the $(-k^2, k_o)$ plane (Fig. 1.2). In the case of elastic scattering, the energy W reduces to the mass of the initial nucleon, W = m, so that (1.6) becomes

$$-k^2 = 2m_1k_o$$

and the corresponding straight line passes through the origin.

The threshold of pion electroproduction corresponds finally to

$$W = m_2 + m_\pi, \tag{1.7}$$

so that the straight line (1.6) becomes

$$-k^2 = 2m_1k_0 - (2m_2 + m_\pi)\, m_\pi + m_1^2 - m_2^2.$$

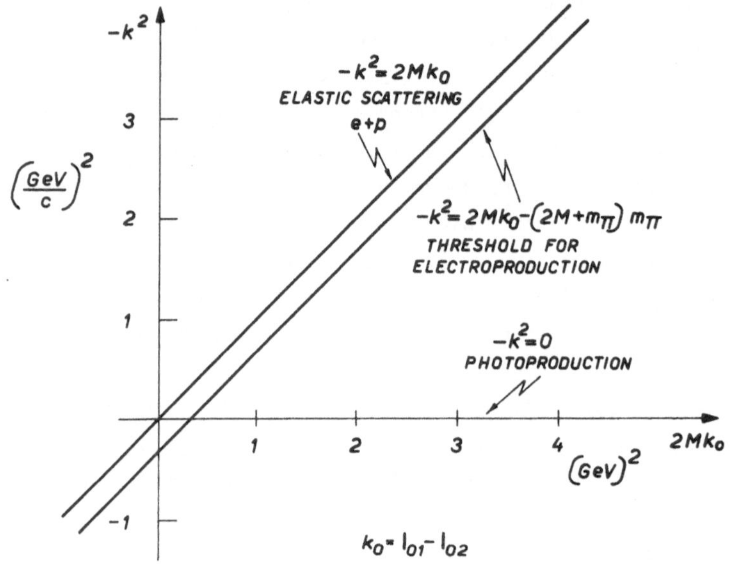

Fig. 1.2. Kinematical relationship between elastic scattering, electroproduction, and photoproduction

1.2 Currents and Hadrons /15/

Let us consider for a moment the general process of (elastic or inelastic) scattering of electrons on a nucleon target

$$e + N \to e' + A. \tag{1.8}$$

where A represents any hadronic state. In the one-photon approximation the amplitude for this process is

$$T = \bar{u}(l_2)\, \gamma^\mu\, u(l_1)\, \frac{e^2}{k^2}\, M_\mu, \tag{1.9}$$

where $\bar{u}(l_2)\, \gamma^{\mu}\, u(l_1)$ is the exactly known (at this order in e^2) form of the lepton electromagnetic vertex, while

$$M_{\mu} = \langle A|\ V_{\mu}^{em}\ |N\rangle \qquad (1.10)$$

is the matrix element of the electromagnetic current between hadronic physical states. This quantity embodies all dependence on strong interactions.

The interest of lepton scattering on hadrons is that the electromagnetic and weak currents have been understood, in the framework of current algebra, to be among the fundamental observables of quantum field theory. As a consequence, the experimental knowledge of their matrix elements is of great importance for our understanding of hadron physics.

It is well known that there is a close analogy between the different kinds of currents, electromagnetic and weak: they are all of vector or pseudovector character and they all obey exact or approximate conservation laws. Moreover, the existence of exact commutation laws among the corresponding charges has led to several theoretical predictions in good agreement with experiment. A particularly interesting application of these ideas is that the electroproduction of a single pion near threshold is directly connected with the matrix element of the axial current between nucleon states.

If one looks at the different kinds of final states which can be produced in electron-nucleon scattering, the simplest and most basic process is of course elastic scattering, which leads directly to the study of the electromagnetic form factors of the target particle.

The possibilities afforded by elastic scattering experiments are limited by the fact that the only "clean" target is the proton target; neutron form factors must be extracted from the data obtained in electron-deuteron scattering.

One of the important roles of pion electroproduction at low energy is to provide information about electromagnetic form factors which cannot be obtained simply by direct elastic scattering. These are: a) the neutron form factors, b) the pion form factor, and c) the nucleon-$\Delta(3,3)$ transition form factors. As shown in Fig. 1.3, they all appear in appropriate one-particle contributions to the electroproduction amplitude. We shall indeed show that experimental comparison of the theoretical electroproduction amplitude leads to reasonable estimates of some of these form factors.

Another important application of low-energy electroproduction, to be discussed in detail in this review, is the determination, via a current algebra low-energy theorem, of the axial form factor of the nucleon.

Let us finally recall that inelastic electron scattering is becoming more and more important as a probe of hadron structure. This is seen if one moves from the

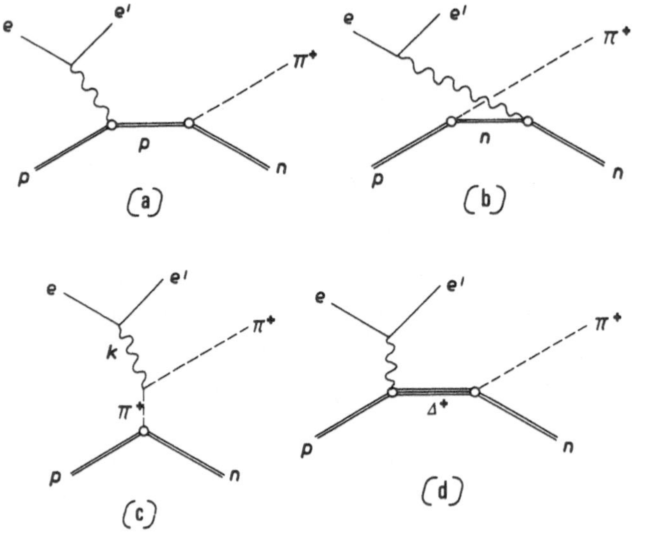

Fig. 1.3. Feynman diagrams that dominate low-energy electroproduction in the o.p.e.a.: (a-c) Born terms, (d) isobar term

low-energy inelastic region to deep inelastic scattering in correspondence to large values of both k^2 and the "mass" of the final hadronic states (W). These experiments do indeed lead to fundamental information about the possible elementary components of the nucleon, their electromagnetic and weak interactions, and their distribution inside hadrons.

1.3 The Electroproduction Cross Section

In the o.p.e.a., the electroproduction cross section can conveniently be expressed in the form /14/ (Appendix B)

$$\frac{d^5\sigma}{dl_{02}\,d\Omega_1\,d\Omega_\pi^*} = \Gamma_t \frac{d\sigma_v}{d\Omega_\pi^*}, \tag{1.11}$$

where the first factor

$$\Gamma_t = \frac{\alpha}{2\pi^2}\frac{l_{02}}{l_{01}}\frac{k_L}{-k^2}\frac{1}{1-\varepsilon} = \frac{\alpha}{\pi^2}\frac{l_{02}}{l_{01}}\frac{W^2 - m_N^2}{2m_N(-k^2)}\;\frac{1}{1-\varepsilon} \tag{1.12}$$

is of electromagnetic origin and contains the effect of the electron-photon vertex and the photon propagator. It can be interpreted as the number of virtual photons per scattered electron in dl_{02} and $d\Omega_1$. In (1.12), k_L is the energy that a real photon must have in the l.f. for producing the final system (πN) with the same invariant mass W.

The second factor in (1.11) is the c.m. differential cross section for pion photoproduction by virtual monochromatic polarized photons. It has the general form

$$\frac{d\sigma_V}{d\Omega_\pi^*}(W, k^2, \varepsilon, \theta_\pi^*, \phi_\pi^*) = A + \varepsilon B + \varepsilon C \sin^2\theta_\pi^* \cos 2\phi_\pi^* + \sqrt{\varepsilon(1+\varepsilon)}\, D \sin\theta_\pi^* \cos\phi_\pi^*, \tag{1.13}$$

where A, B, C, and D are structure functions, related to $<N_1|V_\mu|\pi N_2><\pi N_2|V_\nu|N_1>$, which depend only on the variables W, k^2, t (or θ_π^*) but not on ε and ϕ_π^*. The meaning of the four terms is very simple. A is the differential cross section for unpolarized transversal virtual photons. In the limit $k^2 \to 0$ it approaches the photoproduction cross section by real photons. The term εB is the differential cross section due to longitudinal photons; $\varepsilon C \sin^2\theta_\pi^* \cos 2\phi_\pi^*$ is the modification of the cross section due to transverse linear polarization, as shown by the factor ε and its dependence on ϕ_π^*. Finally the term containing D originates from the interference between longitudinal and transverse components. In the limit $k^2 \to 0$, B and D vanish and $d\sigma_V/d\Omega_\pi^*$ goes over into the well-known photoproduction cross section for linearly polarized photons of polarization ε.[1]

In (1.11)-(1.13) $d\Omega_1$, l_{01}, l_{02}, and $|\underline{k}|$ are measured in the l.f., $d\sigma_V/d\Omega_\pi^*$ in the frame of the c.m. of the (πN) system.

The terms A and C can also be written in the form

$$A \equiv \sigma_u = \frac{1}{2}(\sigma_\parallel + \sigma_\perp),$$
$$C \sin^2\theta^* \equiv \sigma_p = \frac{1}{2}(\sigma_\parallel - \sigma_\perp), \tag{1.14}$$

where $\sigma_\parallel$ and $\sigma_\perp$ are the cross sections for transverse photons polarized parallel and perpendicular to the production plane. The term containing $D \equiv \sigma_I$ takes account of the interference between the components $\sigma_\parallel$ and σ_T of the virtual photon polarization.

By integrating (1.11) with respect to the direction of emission of the pion, we obtain /16/

$$\frac{d\sigma}{dl_{02}\, d\Omega_1} = \Gamma_t \left[\sigma_T(W, k^2) + \varepsilon\, \sigma_L(W, k^2)\right], \tag{1.15}$$

where

---

1 The cos $2\phi_\pi^*$ term also occurs in experiments with polarized bremsstrahlung beams. The cos ϕ_π^* term is unique in electroproduction.

$$\sigma_T = \int A\ (W,\ k^2,\ \theta_\pi^*)\ d\Omega_\pi^*,\quad \sigma_L = \int B\ (W,\ k^2,\ \theta_\pi^*)\ d\Omega_\pi^*, \tag{1.16}$$

are the so-called longitudinal and transversal cross sections of virtual photons.

These are the quantities that are measured in single-arm experiments. By taking measurements for different values of ε while W and k^2 are kept constant, one can separate σ_L from σ_T, while the terms C and D can be obtained only from coincidence experiments in which the experimenter takes advantage of their characteristic azimuthal dependence.

In conclusion, (1.11)-(1.16) summarize all our knowledge of the electroproduction cross section based on quantum electrodynamics in the o.p.e.a. The physics in which we are interested is completely contained in the four structure functions A, B, C, D and the two integral cross sections (1.16).

Under the assumption that only s and p waves in the (πN) system contribute to the cross section (1.13), the structure functions can be expanded as follows in terms of *angular coefficients:*

$$\begin{aligned} A &= A_0 + A_1 \cos\theta_\pi^* + A_2 \cos^2\theta_\pi^*, \\ B &= B_0 + B_1 \cos\theta_\pi^* + B_2 \cos^2\theta_\pi^*, \\ C &= C_o, \\ D &= D_o + D_1 \cos\theta_\pi^*. \end{aligned} \tag{1.17}$$

If also d(f) waves are taken into account, the following angular coefficients should be added: A_3 (A_4), B_3 (B_4), C_1 (C_2), D_2 (D_3).

Whenever the measurements are taken at a single value of the scattering angle θ_1, the structure functions A and B cannot be separated and the quantities that can be derived from the analysis of the observed angular distribution are

$$\begin{aligned} &\bar{A}_0 = A_0 + \varepsilon B_0,\quad \bar{A}_1 = A_1 + \varepsilon B_1, \\ &\bar{A}_2 = A_2 + \varepsilon B_2,\ C_0,\ D_0,\ D_1. \end{aligned} \tag{1.18}$$

Finally, the angular coefficients appearing in (1.17) may be decomposed into multipole amplitudes as shown in Appendix C.

2. Quantities of Physical Interest

The electroproduction process has a rather unique role in the sense that it offers the possibility of correlating and measuring, in particular kinematical conditions, some interesting parameters of the hadron world. We have in mind, in particular, the nucleon and pion form factors. The problem is somewhat intrigued by a series of features, such as gauge invariance and approximate chiral symmetry, which make it interesting but complicated at the same time, since alternative, even if complementary, descriptions are available.

Grossly speaking, the initial idea is to study the phenomenon for configurations such that a particular contribution is reasonably expected to dominate. Unfortunately this occurs in most cases for values of the kinematical variables which do not belong to the physical region, even if they are out by small quantities; more precisely, the parameter playing the main role in all these considerations is the pion mass, much smaller than any other mass in the hadron world. In particular, for the case we are discussing, the significant ratio should be $m^2_\pi/m^2_N \approx 0.02$.

Consider the structure of the singularities of the electroproduction amplitude. The nucleon pole, whose residue is expressed in terms of the nucleon electromagnetic form factors, is located at $s = m^2_N$, not so far from the physical region $s > (m_N+m_\pi)^2$, and one can expect the behavior at threshold to be strongly influenced by the nucleon term. Since at threshold $t = (k^2-m^2_\pi)/(1+m_\pi/m_N)$, the pion pole (at $t = m^2_\pi$) must also contribute in some way. Actually the relevant residues, the nucleon and the pion electromagnetic form factors, are correlated as a consequence of electromagnetic current conservation, which leads to a consistency relation.

It is possible, however, to devise an ad hoc phenomenological and automatically gauge-invariant form of the electroproduction amplitude, where only the polar diagrams are suitably included. One then expects, according to the previous considerations, that such a "generalized Born approximation" can represent a reasonable description of electroproduction in the threshold region. In particular, the pion pole contribution dominates the longitudinal charged production. As an outcome, comparison with experiments can allow the determination of the neutron and pion form factors which, as we shall learn soon, are not unambiguously measured in elastic scattering processes. Of course these intuitive arguments have to be supported by a careful estimate of the higher state contributions, to be evaluated with the techniques discussed in detail in Sections 3.2 and 3.3.

More directly, one can consider the possibility of performing experiments in a kinematical region where the pion exchange diagram dominates, or even selecting the pion pole residue by means of more or less refined extrapolation techniques. All these procedures introduce some ambiguity in the determination of $F_\pi(k^2)$.

There is, on the other hand, another contribution to the amplitude which can be viewed as representative of the whole class of diagrams corresponding to the exchange in the t-channel of states with $I^G = 1^-$, $J^{PC} = 1^{++}$ [the A_1(?) and so on]. As already mentioned, this contribution, which is better summarized as the axial nucleon form factor, is expected to dominate near threshold owing to a different mechanism, still based, however, on the small mass of the pion.

The reason for this lies in the fact that the process we are considering can be related to the time-ordered action of two currents - the electromagnetic and the axial vector (as representative of the pion) ones - on the final and initial nucleon states. The fact of starting with these off-shell quantities allows us to derive a representation for the electroproduction amplitude, which involves, besides the electromagnetic, also the weak nucleon vertex. Thus measuring threshold electroproduction (or related processes), one may be able, in principle, to gain independent information on the axial nucleon form factors. Again a theoretical framework is required for a reliable interpretation of the data.

It is clear that the fact of being equipped with different descriptions of the same amplitude leads to consistency conditions, which allow the derivation of sum rules where, besides form factors, a whole continuum of states, in particular the high-energy part, play an important role. This subtle interconnection among different aspects of the phenomenon is one of the interesting points of electroproduction at low energies. For these reasons, most of the theoretical arguments of this review will be devoted to the constraints deriving from gauge and chiral invariance, to their relation with dynamics, and to their observable consequences.

A different situation occurs as far as the first pion-nucleon resonance, the $\Delta(3,3)$, is concerned. This is obviously due to the fact that, working around $s = M_\Delta^2 > (m_N + m_\pi)^2$, one can quite nearly select the Δ contribution so that measurements in the first resonance region lead to a direct determination of interesting parameters describing the electromagnetic transitions between N and Δ. The reason we include this topic here is that the $\Delta(3,3)$ is by now being considered as a natural partner of the nucleon in the quark model and in symmetry schemes like SU_6. Furthermore, the $\Delta(3,3)$ resonance plays an important role in the solution of dispersion relations in the low-energy region and in the saturation of sum rules.

In the following we shall briefly recall the definition of these quantities, whose experimental determination is in principle also available via pion electroproduction at low energies. Other independent, more direct, information on the same quantities will be examined briefly.

2.1 Definitions /17, 18/

We recall that the electromagnetic form factors of a physical system are functions of the momentum square of the virtual photon, k^2, which summarize in a global, phenomenological way the effect of all detailed processes contributing to the photon-target particle interaction. Physically they can be interpreted as the manifestation of the space-time extension of the target-particle viewed as due either to the virtual meson cloud, or to its parton structure, depending on the conceptual frame adopted for the interpretation of the experimental results. This intuitive picture becomes more precise in the "nonrelativistic", static limit $\Delta E = k_0 \to 0$, where the form factors are a direct consequence of the extended structure of the particle and are the Fourier transforms of the charge, magnetic moment and higher moment distributions. (Their number depends of course on the spin of the particle.) For instance, the most frequently used dipole distribution $F(-\underline{k}^2) = (1+\underline{k}^2/\alpha^2)^{-2}$ corresponds to an exponential density $\rho(r) = (8\pi/\alpha^3)e^{-\alpha r}$, and, in this particular framework, the large dependence on k^2 reflects the finite, nonvanishing slope at the coordinate origin.

2.1.1 The Pion Electromagnetic Vertex

The pion electromagnetic form factor is defined by the relation $[k^2 = (p_2 - p_1)^2]$

$$\langle\pi(p_2)\,|V_\mu^{em}|\,\pi(p_1)\rangle = (p_1 + p_2)_\mu\, F_\pi\,(k^2) \tag{2.1}$$

Since the pion has spin zero, only an electric type contribution is present, while a term $\propto k_\mu$ is ruled out by the current conservation condition $\partial^\mu V_\mu = 0$.

$F_\pi(k^2)$ is a Lorentz invariant function of k^2, which, as a consequence of the hermiticity of the electromagnetic current, must be real in the spacelike region $k^2 < 0$ and below the timelike threshold $k^2 < 4m^2_\pi$. It becomes complex for $k^2 > 4m^2_\pi$ (analyticity guarantees we are dealing with the same function). Furthermore, the normalization to the charge value $\langle Q\rangle/\langle e\rangle = 1$ gives the condition[2]

$$F_\pi(0) = 1. \tag{2.2}$$

[2] We adopt the invariant normalization

$$\langle p_2|p_1\rangle = 2E\,(2\pi)^3\,\delta(\underline{p}_2 - \underline{p}_1),$$

$$\bar{u}u = 2m,\; u^+u = 2E.$$

2.1.2 The Nucleon Electromagnetic Vertex

The customary definition in terms of the so-called Dirac form factors $F_{1,2}(k^2)$ is

$$<N(p_2)|V_\mu^{em}|N(p_1)> = \bar{u}(p_2)\left[\gamma_\mu\ F_1(k^2) + i\frac{\sigma_{\mu\nu}k^\nu}{2m_N}\ F_2(k^2)\right]u(p_1),$$
$$\sigma_{\mu\nu} = \frac{i}{2}\ [\gamma_\mu,\ \gamma_\nu] \tag{2.3}$$

(a term in k_μ is again forbidden by general arguments).

In the spacelike region, $F_1(k^2)$ and $F_2(k^2)$ are real for the same reason as above and imaginary parts appear for $k^2 > 4m_\pi^2$.

A more significant separation can be derived working in the Breit frame. The idea is to separate in an invariant way the contribution of the charge density part of the electromagnetic current from the current density, i.e., from the vector part. This is done by noticing that the previous matrix element can be rewritten in the form

$$<N(p_2)|V_\mu^{em}|N(p_1)> = \frac{m_N}{P^2}\ \bar{u}(p_2)\ \left[P_\mu G_E(k^2) + N_\mu\ \frac{G_M(k^2)}{2m_N}\right]\ u(p_1), \tag{2.4}$$

where

$$P_\mu = \frac{1}{2}\ (p_1 + p_2)_\mu,$$
$$N_\mu = i\ \varepsilon_{\mu\nu\lambda\sigma}\ P^\nu\ k^\lambda\ \gamma^\sigma\ \gamma_5\ , \tag{2.5}$$

and

$$G_E(k^2) = F_1(k^2) + (k^2/4m_N^2)\ F_2(k^2),$$
$$G_M(k^2) = F_1(k^2) + F_2(k^2), \tag{2.6}$$

are the well-known Sachs form factors.

This separation is meaningful. In fact, take $\underline{P} = \underline{0}$ ($k_0 = 0$)

$$<\underline{p}\ |V_0|-\underline{p}> = 2m_N\ G_E\ (k^2) \qquad \text{(electric part)}, \tag{2.7}$$

$$<\underline{p}\ |\underline{V}\ |-\underline{p}> = i\ \underline{\sigma}\times\underline{k}\ G_M\ (k^2) \qquad \text{(magnetic part)}. \tag{2.8}$$

In so doing, G_M and G_E are also naturally connected to transitions between states of equal and opposite helicity ($1/2 \rightarrow 1/2$ for G_M, $1/2 \rightarrow -1/2$ for G_E).

As a consequence the cross section for electron-nucleon scattering, the so-called Rosenbluth formula /17, 18/ is diagonal in G_M, G_E, i.e.,

$$\frac{d\sigma}{d\Omega} = \left(\frac{d\sigma}{d\Omega}\right)_{Mott} \left(\frac{G_E^2 - (k^2/4m_N^2)\, G^2{}_M}{1 - k^2/4m^2{}_N} - \frac{k^2}{2m^2{}_N}\, G^2{}_M\, tg^2\theta/2 \right) , \qquad (2.9)$$

where $(d\sigma/d\Omega)_{Mott}$ is the Mott cross section for a structureless nucleon.

If the isospin structure of the electromagnetic current $V_{em} = V^{(s)} + V^{(3)}$ is taken into account, one has for proton and neutron

$$F^{(P,N)} = \frac{1}{2}\,(F^{(s)} \pm F^{(v)}). \qquad (2.10)$$

Then

$$F_1{}^{(s)}(0) = F_1{}^{(v)}(0) = 1,$$

$$F_2{}^{(s)}(0) = k_p + k_n \approx -0.12,$$

$$F_2{}^{(v)}(0) = k_p - k_n \approx 3.70,$$

where $k_p = 1.7928$, $k_n = -1.931$ are the nucleon anomalous magnetic moments (in units $e/2m_N$).

2.1.3 The Nucleon Axial Vertex

The nucleon axial transition is described by the matrix element

$$\langle N(p_2)|A_\mu^\alpha|N(p_1)\rangle = \bar{u}(p_2)\,\frac{\tau^\alpha}{2}\,[\gamma_\mu G_A(k^2) + k_\mu\, G_P(k^2)]\,\gamma_5\, u\,(p_1), \qquad (2.11)$$

where A_μ^α is the axial vector current, transforming as an isospin triplet.

The further independent covariant $\sigma_{\mu\nu}k^\nu\gamma_5$ would be the signal for the presence in the axial weak current of a "second class" part with opposite transformation properties under the so-called G-parity transformation.

Since evidence (from nuclear physics) of such a contribution is still being debated we shall omit it /19/.

Equation (2.11) can be rewritten in the form

$$\langle N(p_2)|A_\mu^\alpha|N(p_1)\rangle = \bar{u}(p_2)\,\frac{\tau^\alpha}{2}\,[(\gamma_\mu - 2m_N\,\frac{k_\mu}{k^2})\, G_A(k^2) + \frac{k_\mu}{k^2}\, D(k^2)]\gamma_5\, u\,(p_1) \qquad (2.12)$$

with

$$D(k^2) = 2m_N\, G_A(k^2) + k^2\, G_p(k^2). \qquad (2.13)$$

The separation of (2.12) distinguishes the longitudinal from the transversal character of the matrix element. This is better seen in the Breit frame, where

$$\langle \underline{p}|A_0|-\underline{p}\rangle = 0, \quad \langle \underline{p}|\underline{A}\cdot\underline{n}|-\underline{p}\rangle = \underline{\sigma}\cdot\underline{n}\, D(k^2),$$
$$\langle \underline{p}|\underline{A}\times\underline{n}|-\underline{p}\rangle = 2E\underline{\sigma}\times\underline{n}\, G_A(k^2), \quad \underline{n} = \underline{k}/|\underline{k}|. \tag{2.14}$$

Correspondingly, only the k^2 channel exchange of spin-one states contributes to $G_A(k^2)$, and of spin-zero to $D(k^2)$, which takes into account also the nonconserved nature of the axial current

$$\langle N(p_2)|\partial^\mu A^\alpha_\mu|N(p_1)\rangle = i\bar{u}(p_2)\frac{\tau^\alpha}{2}\gamma_5\, u(p_1)\, D(k^2). \tag{2.15}$$

We finally have from neutron β-decay the value at $k^2 \simeq 0$

$$G_A(0) = 1.260 \pm 0.012. \tag{2.16}$$

We take advantage of the formal analogy to recall the customary form of the pseudoscalar pion-nucleon coupling, which is

$$\langle N(p_2)\pi^\alpha|N(p_1)\rangle = i\bar{u}(p_2)\gamma_5\tau^\alpha u(p_1) g_{\pi N}. \tag{2.17}$$

Experimentally /20/

$$g_{\pi N} \simeq 13.5. \tag{2.18}$$

2.1.4 The N-Δ Electromagnetic Transition

An approximate description of photo- and electro-production in particular energy regions is obtained assuming that the phenomenon proceeds through the production and the subsequent decay of resonances. Treating these states as particles of nearly zero width, one is naturally led to consider the transition vertex $NN^*\gamma$. Now given a N^* complex of spin $j = l + 1/2$ and parity $(-1)^{l+1}$, there are six possible transitions to the γ-nucleon system, which can be classified, for instance, according to character of the photon, i.e., transverse or longitudinal, and to the angular momentum j. One has, for the complete process, magnetic transitions $M_{l\pm}[L_\gamma = l$, parity $= (-1)^{L_\gamma+1}]$, electric $E_{l\pm}$, and longitudinal $L_{l\pm}[L_\gamma = l+1$, parity $= (-1)^{L_\gamma}]$ transitions. Correspondingly it is convenient to use for the $NN^*\gamma$ vertex a form factor decomposition which describes physical, i.e. helicity or multipole, transitions in a given reference frame, thus leading to "diagonalized" expressions for the cross section.

In other words, we can look for the analogues of both the Sachs form factors G_E, G_M of the nucleon and of the (simpler) Dirac ones F_1 and F_2. This problem has been discussed in some detail in the literature and we reproduce here only the main results concerning the case I=1, in particular the low-energy (3,3) resonance which is of interest to us /21/.

Its properties and quantum numbers are

$$M_\Delta = 1232 \text{ MeV}, \ \Gamma_\Delta = 110 \div 120 \text{ MeV}, \ j^P = \frac{3}{2}^+, \ I = \frac{3}{2}. \tag{2.19}$$

Correspondingly there are the magnetic dipole M_{1+}, electric, and Coulomb quadrupole E_{1+}, L_{1+} transitions.[3]

One then defines a set of form factors which can be shown to be free of kinematical singularities and constraints, thus representing a convenient framework for the discussion of theoretical models /22/

$$<\Delta(p^*)|V_\mu|N(p)> = \Gamma_{1\mu}\, G_1(k^2) + \Gamma_{2\mu}\, G_2(k^2) + \Gamma_{3\mu}\, G_3(k^2), \tag{2.20}$$

with

$$\begin{aligned}
\Gamma_{1\mu} &= \bar{u}^\nu(p^*)(k_\nu\gamma_\mu - \gamma\cdot k\, g_{\mu\nu})\gamma_5\, u(p),\\
\Gamma_{2\mu} &= \bar{u}^\nu(p^*)(k_\nu P_\mu - P\cdot k\, g_{\mu\nu})\gamma_5\, u(p),\\
\Gamma_{3\mu} &= \bar{u}^\nu(p^*)(k_\nu k_\mu - k^2\, g_{\mu\nu})\, \gamma_5\, u(p).
\end{aligned} \tag{2.21}$$

Conversely one can introduce form factors, which are directly related to physical transitions and therefore useful for experimental analysis. These are defined as follows:

$$<\Delta(p^*)|V_\mu|N(p)> = M_\mu\, G_M^{(1)}(k^2) + E_\mu\, G_E^{(1)}(k^2) + C_\mu\, G_C^{(1)}(k^2), \tag{2.22}$$

where the magnetic, electric, and Coulomb covariants M_μ, E_μ, C_μ are

$$\begin{aligned}
M_\mu &= a\, \varepsilon_{\mu\nu\lambda\rho}\, \bar{u}^\nu(p^*)P^\lambda k^\rho\, u(p),\\
E_\mu &= -M_\mu + b\, \varepsilon_{\mu\nu\lambda\rho}\, M^\nu P^\lambda k^\rho \gamma_5,\\
C_\mu &= \frac{b}{2}\, (\bar{u}^\nu k_\nu)(k^2 P_\mu - P\cdot k\, k_\mu)\, \gamma_5\, u(p).
\end{aligned} \tag{2.23}$$

---

[3] The nomenclature refers to $L_\gamma = 1,2$, respectively.

The relationship between the form factors of the two sets can easily be found and we do not reproduce it here. In these formulae $u^{\nu}(p^*)$ is the Rarita-Schwinger wave function for the 3/2 particle, $k = p^*-p$, $P = \frac{1}{2}(p^*+p)$, and the kinematical factors a, b, introduced for convenience reasons, are

$$a = -\frac{3}{2}\,\frac{m_N + M}{m_N}\,\frac{1}{(m_N+M_\Delta)^2 - k^2}\,,\quad b = -\frac{4a}{(m_N-M_\Delta)^2 - k^2}\,.$$

The main advantage of this, at first sight complicated, decomposition is the direct proportionality of the form factors $G_M^{(1)}$, $G_E^{(1)}$, $G_C^{(1)}$ to the corresponding multipoles. The proportionality factor is, of course, energy dependent and takes into account in particular the finite width effects [e.g., $M_{1+} \propto G_M^{(1)}(k^2)\,(s - M_\Delta^2 + i\,M_\Delta\Gamma)^{-1}$]. We also mention the simple, diagonal, form of the electroproduction cross section at the resonance, $s \Rightarrow M^2$. One finds that

$$\sigma_T + \varepsilon\sigma_L = \frac{4m_N}{\Gamma M_\Delta}\,\frac{\alpha|\underline{k}_L|^2}{M_\Delta^2 - m_N^2}\left[1 - k^2/(M_\Delta + m_N)^2\right]^{-1}$$
$$\left[G_M^{(1)\,2} + 3\,G_E^{(1)\,2} - (k^2/m_N^2)\varepsilon\,G_C^{(1)\,2}\right]. \tag{2.24}$$

The experimental determination of the N-Δ transition form factors will be discussed later. As an indication we anticipate the $k^2 = 0$ values obtained from photoproduction (where $G_C^{(1)}$ does not contribute)

$$G_M^{(1)}(0) = 2.74 \div 3.00,$$
$$G_E^{(1)}(0) = 0.03 \div 0.12. \tag{2.25}$$

For use later on we give also the form of the $N\Delta\pi$ vertex. This is

$$\langle N(p)\pi^{\alpha}(q)|\Delta(p^*)\rangle = \frac{g^*}{m_N}\,\bar{u}(p)q^{\mu}u_{\mu}^{\alpha}(p^*), \tag{2.26}$$

and from the experimental width one has /20/

$$g^{*2}/4\pi \approx 15. \tag{2.27}$$

2.1.5 The N-Δ Axial Vector Transition

The matrix elements of the axial vector current between a nucleon and its excitations are of direct relevance for a description of the neutrino production of pions

$$\nu_\mu + N \to \mu^- + N' + \pi.$$

The structure of these matrix elements is analogous to the vector transition case (of course taking into account the opposite parity character), with the additional contribution due to the pseudoscalar component of the axial current; its presence is a consequence of the fact that $\partial^\mu A_\mu$ is in general not vanishing. In other words, besides the transverse and longitudinal transitions, $\mathcal{M}_{1\pm}$, $\mathcal{E}_{1\pm}$, and $\mathcal{L}_{1\pm}$, there will now be scalar multipoles $\mathcal{H}_{1\pm}$ related to $\langle\Delta|\partial^\mu A_\mu|N\rangle$. In particular, for the $N \overset{A}{\to} \Delta$ transition the meaningful quantities are $\mathcal{M}_{1+}$, $\mathcal{E}_{1+}$, $\mathcal{L}_{1+}$, $\mathcal{H}_{1+}$ /23/.

We are not directly concerned with the neutrino-production process. However, as for the nucleon, the $\langle\Delta|A_\mu|N\rangle$ vertex will be introduced in a description of $\pi\Delta$ electroproduction $e + N \to e' + \pi + \Delta$, based on current algebra equal time commutator. This topic will be touched in the final part of this book, and for that purpose we shall use the following simple (theoretically) decomposition /24/:

$$\langle\Delta(p^*)|A_\mu|N(p)\rangle = \bar{\Gamma}_{1\mu}H_1 + \bar{\Gamma}_{2\mu}H_2 + \bar{\Gamma}_{3\mu}H_3 + \bar{\Gamma}_{4\mu}H_4, \tag{2.28}$$

where

$$\begin{aligned}
\bar{\Gamma}_{1\mu} &= \frac{1}{M}\,\bar{u}^\nu(p^*)(k_\nu\gamma_\mu - \gamma\cdot k\, g_{\mu\nu})\, u(p), \\
\bar{\Gamma}_{2\mu} &= \frac{1}{M^2}\,\bar{u}^\nu(p^*)\,(k_\nu P_\mu - P\cdot k\, g_{\mu\nu})\, u(p), \\
\bar{\Gamma}_{3\mu} &= \bar{u}^\nu(p^*)\,(k_\nu k_\mu/k^2 - g_{\mu\nu})\, u(p), \\
\bar{\Gamma}_{4\mu} &= \bar{u}^\nu(p^*)\,\frac{k_\nu k_\mu}{k^2}\, u(p),
\end{aligned} \tag{2.29}$$

and the $H_i(k^2)$'s are a set of form factors with good analyticity properties. Notice that the above definition is such that

$$\langle\Delta(p^*)|\partial^\nu A_\nu|N(p)\rangle = H_4(k^2)\,\bar{u}^\nu(p^*)\,k_\nu u(p). \tag{2.30}$$

New data on the reaction $\nu p \to \mu^- p\pi^+$ have been produced recently and we shall summarize them in Section 2.2.3.

2.2 Possible Sources of Information [4]

No evidence has been found so far of a structure of charged leptons. Therefore the extreme assumption that they are *pointlike Dirac particles* is acceptable and

---

4 We recall that, in $\hbar=c=1$ units, 1 fm^{-1} = 197.32 MeV, i.e., $1(GeV)^2 = 25.69\ fm^{-2}$. Also 1 $fm^{-2} \approx 2\, m_\pi^2$. In the following there will be an (innocent) inconsistency in our notations: masses will be measured in MeV but for momenta MeV/c (or GeV/c) are used.

represents the first basic hypothesis underlying present quantum electrodynamics. The second basic assumption of this theory is that the *photon propagator* is k^{-2}. The limits of validity of these two assumptions are reexamined periodically and can be found elsewhere /25/.

We only recall that $e^- - e^-$ and $e^+ - e^-$ scattering experiments show that the form factor of the electron is found equal to 1 to within 1 or 2% up to $k^2 \lesssim 1.5$ $(GeV/c)^2$ and within $\approx$ 5% up to about 2.5 $(GeV/c)^2$ /26/.

The *validity of the o.p.e.a.* is the third underlying assumption of wide chapters of particle physics, in particular of all the problems reviewed in the present book.

From the first basic assumption it follows that charged leptons, i.e., electrons and muons, provide excellent (or at least very satisfactory) probes of the structure (e.m. as well as w.) of hadrons, in particular of stable hadrons such as protons and neutrons.

Most of the experimental work has been done with electrons which are available in intensive, well collimated, monoenergetic beams, the energy of which can easily be changed. So far the use of muons has been rare and mainly devoted to explore whether some difference in their behaviour could be found with respect to electrons, apart from the much greater value of the mass.

Many excellent review articles have appeared, also recently, on the elastic scattering of electrons and muons /26/ or of neutrinos /27/ on nucleons or on the production of hadron-antihadron pairs by e^+e^- annihilation /28, 29/. In the present book, devoted to the information obtained on hadron form factors from inelastic scattering of leptons, only some of the results obtained from elastic scattering experiments or from e^+e^- annihilation /30/ are summarized in this section. They provide the background and the frame within which the main subject of the present article should be viewed.

The validity of o.p.e.a. has been very extensively tested by verifying the linearity of the Rosenbluth plot for both ep and μp scattering as well as by looking for effects determined by the two-photon amplitude, such as a possible difference between the cross section for elastic l^+p and l^-p scattering and a possible polarization of the protons recoiling from lp elastic collisions.

From all these experiments the general conclusion can be drawn that the contribution to the elastic cross section originating from two-photon exchange processes does not exceed a few percent up to the value of $(-k^2)$ of the order of at least 5 $(GeV/c)^2$.

It is also interesting to mention the fact that muon beams are produced by a few high-energy machines, such as at SLAC, Brookhaven, and CERN-PS. Experiments with these beams have been performed on muon-proton elastic and deep inelastic scattering with the aim of testing muon-electron universality. A detailed discussion of these results can be found elsewhere. No experiment has been made yet on muon-initiated production of a single pion in the region of or below the first pion-nucleon resonance.

2.2.1 Information on the Pion Form Factor and the Pion Root-Mean-Square Radius

The pion e.m. form factor $F_\pi(k^2)$ has been investigated rather extensively in the time. like region by measuring the cross section of the process $e^+e^- \to \pi^+\pi^-$ /31 - 34/.

We refer the reader to other review articles /4, 26, 29/ for the analysis of the experimental data which clearly show the dominance of the production of the vector meson ρ. The data obtained in the region $k^2<1$ $(GeV/c)^2$ show the expected interference term with the ω contribution. The experimental points obtained in the region $k^2>1$ $(GeV/c)^2$ remain appreciably above the expected contribution of the ρ tail. All these data /31 - 34/ are shown in Fig. 6.3 in connection with the discussion of a few points obtained from inverse electroproduction for $k^2 \gtrsim 0$.

In the spacelike region $k^2 < 0$, elastic scattering $e\pi^\pm$ experiments would clearly provide the most direct method, since, according to (2.1), the cross section of a spin-zero particle is given by

$$\frac{d\sigma}{d\Omega} = \left(\frac{d\sigma}{d\Omega}\right)_{Mott} |F_\pi(k^2)|^2 . \tag{2.31}$$

This type of experiment, however, is impossible because we do not dispose of free pions in the form of targets or of beams of sufficiently high density to provide statistically significant data when crossed by intensive electron beams. Thus, other approaches have been explored. The most extensive results have been obtained from electroproduction and will be discussed in Section 5.2.

Two alternative methods have been considered. Both, however, always involve very low values of $-k^2$ so that only the root-mean-square radius $\langle r_\pi^2\rangle^{1/2}$ of the pion can be deduced where

$$\langle r_\pi^2\rangle = 6(dF_\pi/dk^2)_{k^2=0} . \tag{2.32}$$

The first method is based on the observation of knock-on electrons produced by charged pions. Two rather old experiments using nuclear emulsion exposed to pion beams /35/ allowed only the derivation of not very significant upper limits for $\langle r_\pi^2\rangle^{1/2}$ (~3÷4 fm) (Table 2.1). The only accurate determinations of the pion mean radius by this method have been obtained by two groups. The first one was a Dubna-Los Angeles collaboration /37/ in which, by means of a narrow angle magnetostrictive spark chamber spectrometer, the scattering of negative pions of 50 GeV/c from stationary electrons was observed. The other was a Los Angeles, Notre Dame, Pittsburg,

Table 2.1 The pion root-mean-square radius

Authors	$<r^2>^{1/2}$ fm.	Remarks
	From knock-on electrons produced by pions $k^2 < 0$	
ALLAN et al. /35/	4.5	
CASSEL /36/	3	
SHEPARD et al. /36/	0.9	
ADYLOV et al. /37/	0.7 ± 0.9	p_{π^-} = 50 GeV/c
DALLY et al. /38/	0.56 ± 0.04	p_{π^-} = 100 GeV/c $0.03 \leq -k^2 \leq 0.07$ $(GeV/c)^2$
	From $(\pi^+ + He)/(\pi^- + He)$ scattering comparison $k^2 < 0$	
BLOCK et al. /41/	0.9	Liquid He bubble chamber
CROWE et al. /42/	2.96 ± 0.43	Counter hodoscope Original analysis
CROWE et al. /42/	0.80 ± 0.40	Some data analyzed by NICHITIU /44/
Dubna-Turin collaboration /46/	0.80 ± 0.17	High-pressure He: streamer chamber + counter hodoscope
	From electroproduction (Section 4.5.3) $k^2 < 0$	
AKERLOF et al. /14/	0.80 ± 0.10	
MISTRETTA et al. /47/	0.86 ± 0.14	
BEBECK et al. /48/	0.70 ± 0.007	Computed from best bit
Saclay group /156/	$0.74^{+0.11}_{-0.13}$	single-pole distribution
	From inverse electroproduction (Section 6.1) $k^2 \geq 0$	
DEVONS et al. /49/	< 1.9	$T_{\pi^-} = 0$
BEREZHNEV et al. /50/	0.75 ± 0.14	T_{π^-} = 275 MeV

Batavia, Dubna collaboration /38/, which, at F.N.A.L., measured with a set of proportional wire chambers the knock-on electrons of negative pion of p_{π^-} = 100 GeV/c with a four momentum transfer k^2 between 0.03 and 0.07 $(GeV/c)^2$.

The second method /39, 40/ is based on the comparison of the elastic scattering of positive and negative pions by ^4He.

The analysis of the experimental data requires a detailed knowledge of the interaction of pions with nuclei /40/, which was not available when the first experiments were carried out /41, 42/.

Various improvements were later introduced in the analysis /43, 44/. A valuable review of the subject has been given by NICHITIU and SHCHERBAKOV /45/, who show that the main problem is that the ambiguity in phase shift analysis has a much larger effect on the pion radius than all other Coulomb corrections. They succeeded in finding a phase shift solution which seems to eliminate this difficulty. The most reliable value is that obtained by the Dubna-Torino collaboration /46/. The values obtained by previous authors are given in Table 2.1, mainly to show the development of our knowledge on this interesting subject.

In the same table we anticipate a few values deduced from experiments on electroproduction of π^+ (Section 5.2) and inverse electroproduction ($k^2 \gtrsim 0$) (Section 6.1).

2.2.2 Information on the Electromagnetic Form Factors of the Nucleon

The electromagnetic form factors $G_E^p(k^2)$ and $G_M^p(k^2)$ of the proton have been determined with rather good accuracy in the spacelike region ($k^2 < 0$) up to $-k^2 = 25\ (\mathrm{GeV/c})^2$ from electron-proton elastic scattering experiments /17, 26/.

The experimental points are interpolated roughly by the *dipole formula*,

$$G_E^p(k^2) = (1-k^2/M_V^2)^{-2}, \quad M_V \simeq 0.84\ \mathrm{GeV}, \tag{2.33}$$

and by the *scaling law*

$$G_M^p(k^2) = \mu_p G_E^p(k^2), \quad \mu_p = 1+\kappa_p = 2.79, \tag{2.34}$$

both of which, however, have no theoretical foundation yet and should be considered essentially as convenient rules, the validity limits of which can be found elsewhere /26/. Similarly, a unique explanation for the rapid fall-off of the form factor for increasing $|k^2|$ has not yet been found. Among the various proposals we mention the recent so-called quark-counting rule /51/, based on simple quark model considerations, which predicts for the electromagnetic form factor of a hadron the asymptotic behaviour

$$F_H \sim (k^2)^{1-n_H}$$

(apart from corrections $\propto \ln k^2$), where n_H is the minimum number of constituent quark fields. Thus

$$F_\pi(k^2) \sim (k^2)^{-1}, \quad G_E \sim G_M \sim (k^2)^{-2}.$$

Information in the timelike region $k^2 > 0$ can be derived for

$$k^2 \geq 4m_N^2$$

from production of $p\bar{p}$ pairs by e^+e^- annihilation.

The cross section of the process, however, is so small that with the luminosity of present electron positron machines, only one point has been measured at $k^2 = 4.4\ (GeV/c)^2$ /52/, while from the inverse process ($\bar{p}p \rightarrow e^+e^-$) two values have been obtained at threshold [$\bar{p}$ at rest: $k^2 = 3.52\ (GeV/c)^2$] /53/ and upper limits at $k^2 =$ 5.2 and 6.7 $(GeV/c)^2$ /54/.

Information on e-n elastic scattering is extracted from electron-deuteron scattering once the proton form factors and the structure of the deuteron are known. The results are model dependent since the analysis procedure involves the wave function of the deuteron and its relativistic corrections. For the magnetic form factor of the neutron the scaling law

$$G_M^n(k^2)/\mu_n = G_M^p(k^2)/\mu_p \tag{2.35}$$

seems to hold approximately. The electric form factor of the neutron G_E^n is still known with considerable uncertainty. The value $(dG_E^n/dk^2)_{k^2=0}$ has been derived from the scattering of low-energy (thermal) neutrons on high Z atoms. The results of various experiments are in fair agreement with each other. Their mean value is /55/

$$\left(\frac{dG_E^n}{dk^2}\right)_{k^2=0} = \left(\frac{dF_1^n}{dk^2}\right)_{k^2=0} + \frac{F_2^n(0)}{4m_N^2} = -0.0201 \pm 0.0005\ \text{fm}^2. \tag{2.36}$$

For $k^2 < 0$, information on G_E^n is extracted completely from eD scattering as mentioned above.

Figure 2.1 shows the quantity G_E^n derived by various authors from elastic eD scattering experiments using the Feshbach-Lomon wave functions /26/. The dash dotted line is $G_E^n = \tau\mu_n G_E^p$ corresponding to $F_1^n(k^2) = 0$. The dashed line is

$$G_e^n(k^2) = \mu_n \frac{\tau}{1+4\tau} G_e^p(k^2), \quad \tau = -k^2/4m_N^2.$$

The solid line is best fit to the data points with the curve

$$G_E^n(k^2) = \mu_n \frac{\tau}{1+b\tau} G_e^p(k^2)$$

where b is the free parameter, which turns out to be ≈ 5.6.

At higher values of $|k^2|$ ($> 15\ \text{fm}^{-2}$) the neutron form factor can be determined by means of inelastic electron-deuteron scattering experiments but the result of the analysis is again model dependent.

This situation justifies the attempt, although unsuccessful for the moment, to obtain some further information on $G_E^n(k^2)$ from a different class of experiments (Section 5.1).

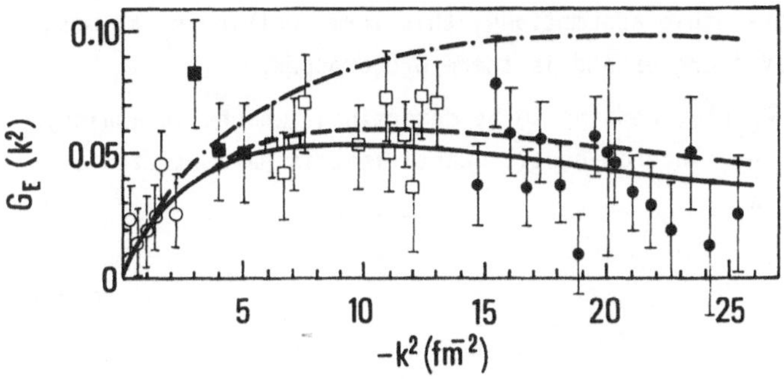

Fig. 2.1. The neutron electric form factor $G_E^n(k^2)$ derived from elastic electron-deuteron scattering measurement. The Feshbach-Lomon wave function was used. The dash dotted line is $G_E^n = \mu_n \tau G_E^p$, which corresponds to the assumption $F_1^n = 0$; the dashed curve is $G_E^n = G_E^p \cdot \mu_n\tau/(1+4\tau)$; the solid curve is the best fit to the data points with a curve $G_E^n = G_E^p\ \mu_n\tau/(1+b\tau)$ (b, the free parameter, turns out to be 5.6). The figure is taken from /26/, to which we refer for the quotations of the original experimental papers

2.2.3 Information on the Axial Vector Form Factors of the Nucleon and of the N-Δ Transition /4, 27/

A) The most direct and natural way of extracting the axial vector form factors $G_A(k^2)$, $G_p(k^2)$ is from the experimental investigation of the neutrino quasi-elastic reactions

$$\nu_\mu n \rightarrow \mu^- p, \quad \bar{\nu}_\mu p \rightarrow \mu^+ n.$$

Taking into account the V-A space-time structure of the weak current $I_\mu^W = V_\mu - A_\mu$, its nucleon matrix element is then expressed in terms of the quantities $\langle N_2|V_\mu|N_1\rangle$, $\langle N_2|A_\mu|N_1\rangle$ already discussed in Sections 2.1.2 and 2.1.3. The standard assumptions, which are made to further simplify the problem, are:

i) The isotriplet current hypothesis (generalized C.V.C.), by which the weak currents belong to the same isotriplet as the electromagnetic one. Then the vector form factors are the same isovector form factors measured in electron-nucleon scattering.

ii) Dominance of the divergence form factor $D(k^2)$ by the pion pole (see Section 3.4 for a more complete discussion). This then gives for the induced pseudoscalar form factor $G_P(k^2)$ the approximate expression, for not too large $|k^2|$,

$$G_P(k^2) \simeq \frac{2f_\pi g_{\pi N}}{m_\pi^2 - k^2} - 2m_N \left(\frac{dG_A}{dk^2}\right)_{k^2=0} . \qquad (2.37)$$

On the other hand, in any reasonable assumptions, this term contributes at most 2-3% to the cross section at GeV energies and is therefore dropped.

Thus one axial form factor, $G_A(k^2)$, remains to be determined, which, in analogy with the observed behaviour of the electromagnetic quantities, is parametrized by a phenomenological dipole formula,

$$G_A(k^2) = G_A(0)\ (1-k^2/M_A^2)^{-2}, \tag{2.38}$$

where

$$G_A(0) = 1.26,$$

as determined from neutron β-decay. Thus the experiment amounts to the determination of the single parameter M_A, which can be extracted from differential and total cross section measurement. The most recent determination of M_A obtained from the data for neutrino scattering on deuterium /56/ gives

$$M_A = 0.95 \pm 0.09\ \text{GeV}, \tag{2.39}$$

and in Figure 2.2 we show the behaviour of the total cross section.

It is interesting to mention at this point the independent information on the weak nucleon form factors one can obtain from the muon capture process in hydrogen,

$$\mu^- p \rightarrow \nu_\mu n,$$

which occurs, for μ-mesons in the 1s orbit, at $k^2 \simeq -0.88\ m_\mu^2$.

The experimental values for Γ_S /57/, the rate for muon capture from singlet $\mu^- p$ state, are in good agreement with the theoretical predictions based on the forms discussed above of the vector and axial vector weak form factors. Conversely, using the world average value of Γ_S and dipole fits for $G_A(k^2)$, $G_E(k^2)$, $G_M(k^2)$, one obtains a range of allowed values for G_P

$$6m_\mu \leq G_P(k^2 = -0.88\ m_\mu^2) \leq 14\ m_\mu. \tag{2.40}$$

[Eq. (2.37) gives $G_P \simeq 8.5\ m_\mu$].

B) The available information on the N-Δ axial transition form factors is derived from πΔ production in neutrino reactions. The k^2 dependence of H_4 is again fixed by requiring the dominance by the pion pole, while for the other form factors the following parametrization is adopted:

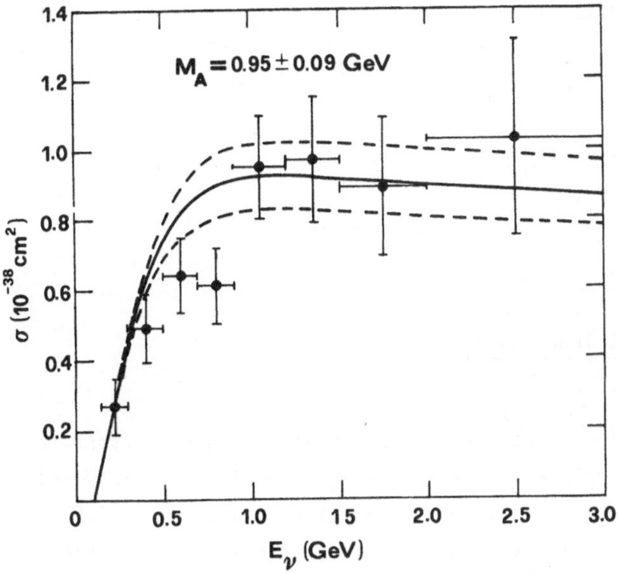

Fig. 2.2. Total cross section for $\nu_\mu n \to \mu^- p$ as measured in the Argonne deuterium experiment with the best fit value for $M_A = 0.95 \pm 0.09$ GeV

$$H_i(k^2) = H_i(0)\left(1 - \frac{k^2}{M_A^{*2}}\right)^{-2}\left(1 - \frac{ak^2}{b-k^2}\right), \tag{2.41}$$

with M_A^*, $H_i(0)$,a,b free parameters to be determined. The fit is of course performed in the framework of a definite theoretical model, and assuming the validity of a model due to Adler, one finds the values /59/

$$H_1(0) \simeq 0, \quad H_2(0) \simeq 0.3, \quad H_3(0) \simeq 1.2,$$
$$a = -1.21\ (\mathrm{GeV/c})^2, \quad b = 2\ (\mathrm{GeV/c})^2. \tag{2.42}$$

while the slope M_A^* turns out to be

$$M_A^* = 0.96\ \mathrm{GeV}. \tag{2.43}$$

not very different from the one obtained for the nucleon axial form factor. Alternative determinations of the form factor $G_P(k^2)$ will be discussed in Sections 5.4.5 (C) and 6.1 (D).

3. Theoretical Approaches

3.1 Introduction: The Theoretical Ingredients

Electroproduction of low-energy pions involves, in an essential manner, strong interactions. As a consequence, a complete set of theoretical predictions would only be possible when a satisfactory theory of strong interactions would be available. We are indeed very far from this ideal situation; however, it has been possible on the basis of general theoretical arguments to make definite and sometimes very precise predictions, which are in good agreement with experimental findings.

A first constraint comes from the unitarity property of the scattering matrix, which of course follows from the general requirement of probability conservation.

For semistrong processes such as electroproduction of hadrons, the relations take the "linear" form

$$\text{Im } M(\gamma N \to \text{Hadrons}) = \sum_{h=\text{int. hadrons}} M^{*}(\gamma N \to h)\ M\ (h \to \text{Hadrons}). \qquad (3.1)$$

Eq. (3.1) can be exploited simply only when the initial energy is so low that the contribution of inelastic channels is negligible. In this case it gives rise to the well-known Fermi-Watson Theorem stating that for each multipole the electroproduction amplitude is a complex quantity whose phase is equal to the corresponding pion-nucleon phase shift.

The next requirements to be discussed are those of Poincaré and SU_2 invariance. They allow us to express the physical amplitude in terms of a fixed number of scalar functions $M_i(\nu,t,k^2)$ (in our case 24) which depend only on the invariants of the problem (in our case the energy-like variable ν, the momentum transfer t, and the virtual mass of the photon k^2).

Further, the extra requirement of gauge invariance leads to linear relations between the fundamental functions M_i, in such a way that we are reduced to only 18 independent functions.

The next theoretical requirements are those of analyticity, whose importance has been stressed in many review articles /59/. Our amplitudes exhibit poles in the kinematical variable in correspondence to single-particle intermediate states. The diagrams corresponding to the nucleon (a and b) and pion (c) poles are shown in Fig. 1.3. A very important property is that the residues of these poles are proportional to the electromagnetic form factors of the nucleon and of the pion.

The full exploitation of analytic properties is made by taking into account, in addition to the one-particle poles, the contributions coming from many-particle states. The general problem involves of course the analytic representation of a function of several complex variables. Although many interesting steps have been made in this direction, most work is done at fixed momentum transfer. This gives the celebrated fixed t dispersion relations.

At this point it is important to note that the combination of gauge invariance and analyticity is not at all trivial. Its offsprings are low-energy theorems expressing exact statements about amplitudes involving zero momentum photons. In our case this leads to the well-known Kroll-Ruderman theorem for the simpler photoproduction process $k^2 = 0$, while for electroproduction the situation is a little more involved.

Let us go back to fixed t dispersion relations whose explicit form is

$$M_i(\nu,t,k^2) = \Big[\text{Nucleon pole}\Big]_i + \frac{1}{\pi}\int \frac{\text{Im}\{M_i(\nu',t,k^2)\}}{\nu'-\nu-i\varepsilon}\,d\nu' . \tag{3.2}$$

Clearly, since we are working at fixed t, and only the singularities in ν have to be selected, the pion pole does not appear explicitly. A question immediately arises: Where has the contribution due to the exchange of a physical pion gone? The answer is not hard to find if one considers that the only extra source of singularities in (3.2) comes from the high-energy behaviour of the dispersion integral. This leads us to consider the high-energy behaviour of Im $M(\nu,t,k^2)$ which is commonly believed to be given in terms of the Regge formula,

$$\text{Im}\,\{M(\nu,t,k^2)\} \underset{\nu\to\infty}{\sim} \sum_j \beta_j(t,k^2)\,\nu^{\alpha_j(t)} , \tag{3.3}$$

where $\alpha_j(t)$ and $\beta_j(t,k^2)$ are the so-called trajectory and residue functions corresponding to the exchange of the j-th Regge pole.

It is readily seen that (3.3) controls the number of subtractions needed in (3.2). At the same time, if one computes the high-energy tail of (3.2) by inserting the Regge pole term (3.3), one sees that poles due to all particles exchanged in the t channel do appear. One thus sees that even if we are interested in low-energy phenomena, it is very hard to exclude high-energy virtual states which give rise to the poles in the t variable, *which are important at low energy*.

On the other hand, if one inserts t channel poles by brute force, one strongly risks *the serious sin of counting the same effect twice*. This last point is a warning against too ambitious attempts. A treatment of our problems in terms of low energy is necessarily incomplete and might lead us to unpleasant compromises.

We finally come to a very important point, which constitutes one of the breakthroughs of particle physics of the last decades /60/.

The pion field is closely associated with axial weak currents (PCAC). This has led to the well-known Goldberger-Treiman relation between the pion-nucleon coupling constant and the pion decay rate.

On the other hand, assuming commutation relations between weak charges, it has been possible to obtain general theorems concerning low-energy pions. This has led to a clear understanding of why the pion s-wave amplitudes are so small (through the so-called Adler consistency relation) and also to a very good prediction about their actual value. Also in our case, soft-pion theory works beautifully, leading to precise theorems relating threshold electroproduction to the axial form factors of the nucleon.

As a consequence of this discussion, we see that because of our incomplete understanding of the problem, experimental comparison of theory is not straightforward. One is naturally led to those comparisons for which the theoretical predictive power is stronger.

A first direction is towards an experimental test of low-energy theorems. There the fundamental problem is to find a reliable manner of extrapolating to the real world from the exact low-energy theorems valid for massless pions. Combination of low-energy limits with dispersion relations leads to exact sum rules which, under appropriate approximations for the continuum, lead to useful relations between the strength of different multipoles.

Another important phenomenological direction is to exploit the fact that at low energy the dispersion relations can be considered as dominated by low-lying resonance. If one restricts one's attention to those resonances whose inelastic decay channels are negligible, the extra constraint due to the Watson theorem greatly increases the predictive power of the isobaric model.

In the framework of such an elementary unitarity-analyticity bootstrap programme, it is indeed possible to get simple order of magnitude evaluations for the electromagnetic parameters of the different resonances.

In the theoretical part of this review we concentrate our attention on those topics which are directly connected with the general structure (invariance and conservation laws) of the electroproduction amplitude and which are of particular relevance in the derivation of low-energy theorems and of the dispersion sum rules which are one of their mathematical consequences. A brief discussion on the dynamical approach to low-energy electroproduction is given with the aim of emphasising the main physical points. For more thorough and complete treatments of the subject we refer to the existing excellent reviews /61/.

3.2 General Properties of the Electroproduction Amplitude

In this section we wish to discuss (or rather recall), in more detail, the structure of the pion electroproduction amplitude and to illustrate briefly, in this specific case, some of the general requirements discussed above.

We study the reaction

$$e^-(l_1) + N(p_1) \rightarrow e^-(l_2) + N(p_2) + \pi(q). \tag{3.4}$$

It is convenient to introduce the lepton and nucleon four-momentum transfers,

$$k = l_1 - l_2, \; \Delta = p_2 - p_1 = k-q, \quad p_1 + k = p_2 + q, \tag{3.5}$$

and the quantity

$$P = \frac{1}{2}(p_1 + p_2). \tag{3.6}$$

We shall find it useful to work in terms of the invariant variables

$$\begin{aligned} k^2 &= (l_1 - l_2)^2, \\ t &= \Delta^2 = (p_2 - p_1)^2, \\ \nu &= q\cdot P = k\cdot P, \\ \nu_B &= -\frac{1}{2}\, q\cdot k, \end{aligned} \tag{3.7}$$

and

$$\begin{aligned} s &= (p_1+k)^2 = (p_2+q)^2 = m_N^2 + 2(\nu-\nu_B), \\ \bar{s} &= (p_1-q)^2 = (p_2-k)^2 = m_N^2 - 2(\nu+\nu_B). \end{aligned} \tag{3.8}$$

Explicit expressions of the quantities in different reference frames and their physical ranges are discussed in Appendix A.

In the one-photon exchange approximation, the matrix element can be written in the form

$$M(e_1N_1 \rightarrow e_2N_2\pi) = e^2\bar{u}(l_2)\gamma^\mu u(l_1)\,\frac{1}{k^2}\,\langle N_2\pi|V_\mu^{em}|N_1\rangle, \tag{3.9}$$

and the effort has to be concentrated on the quantity

$$M_\mu = \langle N(p_2)\pi(q) \mid V_\mu^{em} | N(p_1)\rangle. \tag{3.10}$$

which embodies all the interesting information on strong interactions. We thus turn to an analysis of M_μ and begin by discussing its kinematical structure.

A) Let us first dispose of the charge degrees of freedom. The isospin structure of the electromagnetic current is given by the combination of an isoscalar and of the third component of an isovector, i.e.,

$$V_\mu^{em} = V_\mu^{(s)} + V_\mu^{(v)}, \quad M_\mu = M_\mu^{(s)} + M_\mu^{(v)}. \tag{3.11}$$

Introducing isospin wave functions ψ_α for the pion (α = 1,2,3), v_3 for the virtual photon, and χ for the nucleon isospinor, one can write

$$\begin{aligned} M_\mu^{(s)} &= a^{(o)} M_\mu^{(o)}, \\ M_\mu^{(v)} &= a^{(+)} M_\mu^{(+)} + a^{(-)} M_\mu^{(-)}, \end{aligned} \tag{3.12}$$

with the matrix elements in isospin space to be evaluated for each specific process:

$$\begin{aligned} a^{(+)} &= \psi_\alpha^* \chi^+ \chi \delta_{\alpha 3} v_3, \\ a^{(-)} &= \psi_\alpha^* \chi^+ \tfrac{1}{2}[\tau_\alpha, \tau_3] \chi v_3, \\ a^{(o)} &= \psi_\alpha^* \chi^+ \tau_\alpha \chi. \end{aligned} \tag{3.13}$$

As these formulae show, $a^{(-)}$, $a^{(o)}$ (and the relevant amplitudes) describe the exchange in the t channel of a $\pi\gamma$ (or $N\bar{N}$) complex of isospin 1, G-parity = -1,1, $a^{(+)}$ of a complex of isospin 0, G-parity = -1.

As far as the isospin transitions in the s-channel are concerned ($M_\mu^{(I)}$, I = 1/2, 3/2), one can easily show that

$$M_\mu^{(1/2)} = M_\mu^{(+)} + 2M_\mu^{(-)}, \quad M_\mu^{(3/2)} = M_\mu^{(+)} - M_\mu^{(-)}, \tag{3.14}$$

and for the specific observable processes the relations are given in Table 3.1.

B) We now examine the Lorentz structure of the amplitude M_μ. Introducing for convenience the analogue of a polarization vector for the virtual photon ε_μ (it is actually the lepton matrix element of the electromagnetic current), a very convenient expression of the amplitude is

Table 3.1

Reaction	$a^{(+)}$	$a^{(-)}$	$a^{(o)}$	$a^{(1/2)}$	$a^{(3/2)}$
$\gamma p \to \pi^o p$	1	0	1	1/3	2/3
$\gamma n \to \pi^o n$	1	0	-1	1/3	2/3
$\gamma p \to \pi^+ n$	0	$\sqrt{2}$	$\sqrt{2}$	$\frac{\sqrt{2}}{3}$	$-\frac{\sqrt{2}}{3}$
$\gamma n \to \pi^- p$	0	$-\sqrt{2}$	$\sqrt{2}$	$-\frac{\sqrt{2}}{3}$	$\frac{\sqrt{2}}{3}$

$$\varepsilon^{\mu} M^{\alpha}_{\mu} = \sum_{1}^{8}{}_i \bar{u}(p_2) O_i \; u(p_1) M^{\alpha}_i , \tag{3.15}$$

where the eight covariants have been chosen as follows:

$$\begin{aligned}
O_1 &= \gamma_5 \gamma\cdot\varepsilon, & O_5 &= \tfrac{1}{2}\gamma_5 [\gamma\cdot\varepsilon, \gamma\cdot k],\\
O_2 &= \gamma_5 P\cdot\varepsilon, & O_6 &= \gamma_5 \gamma\cdot k \; P\cdot\varepsilon,\\
O_3 &= \gamma_5 q\cdot\varepsilon, & O_7 &= \gamma_5 \gamma\cdot k \; q\cdot\varepsilon,\\
O_4 &= \gamma_5 k\cdot\varepsilon, & O_8 &= \gamma_5 \gamma\cdot k \; k\cdot\varepsilon.
\end{aligned} \tag{3.16}$$

Correspondingly, eight invariant amplitudes have been introduced which depend only on the scalar variables of the problem, i.e. $M_i(\nu,t,k^2)$, and all general requirements will be expressed through these quantities. Of course, each of the invariant amplitudes M^{α}_i is in turn endowed with isospin exactly as the full amplitude [according to (3.12), for a total of 24 quantities, α = +,-,0].

The above choice for the covariant expansion[5] is the most natural, even if not unique, and it is substantially based on simplicity arguments. It has been shown, indeed that, with the choice (3.16), the relevant amplitudes have, in the variables ν (or $s,\bar{s}$) and t, *only* the singularities which correspond to the propagation of physical systems in those channels, namely they are free of the so-called kinematical singularities[6] which do not allow such an interpretation. Such a property can also

---

[5] Note that for the physical process the amplitudes M_4, M_8 have no direct relevance since $O_4 = O_8 = 0$ as a consequence of the current conservation on the lepton side.

[6] Strictly speaking, this has been rigorously proved only for $k^2 = 0$ /62/.

be confirmed by examining the analyticity properties of the amplitudes in the framework of perturbation theory, and we shall check it explicitly working out the polar contributions.

Kinematical singularities arise when one enforces the current conservation condition $\partial^\mu V_\mu^{em} = 0$, namely $k^\mu M_\mu = 0$. As a consequence, the eight amplitudes M_i are not independent, but turn out to be related by two constraint conditions; elimination of two of the amplitudes in terms of the others introduces spurious singularities in the physical variables. We shall discuss this point in more detail later and proceed with the previous choice of the (not independent) amplitudes.

In concluding this section, we recall another general requirement: the so-called "crossing" which entails simple symmetry properties for the various emplitudes. The crossing constraint has the form /59/

$$M_\mu(p_2,q;\, p_1,k) = -M_\mu^*(p_1,\, -q;\, p_2,\, -k), \tag{3.17}$$

and it can be made plausible on the basis of quantum field theoretical considerations. Intuitively, in the present example, it takes into account the behaviour of the amplitude under the exchange of the two identical nucleons participating in the reaction, while the change of sign of q, k is required by four-momentum conservation.

In terms of scalar variables, all this amounts to $\nu \to -\nu$ (i.e., $s \rightleftarrows \bar{s}$), $t \to t$, $k^2 \to k^2$ and, taking into account the previous definitions, we easily find that

$$M_i^{\binom{\mp}{0}}(\nu,t,k^2) = \begin{vmatrix} - \\ + \\ + \end{vmatrix} \eta_i M_i^{\binom{\mp}{0}}(-\nu,t,k^2). \tag{3.18}$$

The values of the η_i's are fixed by the space properties, while the accompanying column takes into account the isospin character of the amplitude. One has

$$\begin{aligned} \eta_i &= 1, \quad i = 2,\, 5,\, 6, \\ \eta_i &= -1, \quad i = 1,\, 3,\, 4,\, 7,\, 8. \end{aligned} \tag{3.19}$$

C) We now go on to examine the analyticity properties of the amplitudes, and for future considerations it is meaningful to select explicitly the contribution due to the lowest singularities, which correspond to the exchange of stable particles, i.e., the nucleon poles at $s = \bar{s} = m_N^2$ ($\nu = \pm\nu_B$) and the pion pole at $t = m_\pi^2$ ($\nu_B = k^2/4$). Using the standard definition of the electromagnetic vertex, one easily finds for the nucleon singularity the result

$$M_1^{(\pm)} = 0, \quad M_4^{(\pm)} = M_8^{(\pm)} = 0,$$

$$M_2^{(\pm)} = -g_{\pi N} F_1^{(v)}(k^2) \frac{1}{2} \left(\frac{1}{\nu_B-\nu} \pm \frac{1}{\nu_B+\nu}\right),$$

$$M_3^{(\pm)} = -g_{\pi N} F_1^{(v)}(k^2) \frac{1}{4} \left(\frac{1}{\nu_B-\nu} \pm \frac{1}{\nu_B+\nu}\right),$$

$$M_5^{(\pm)} = -g_{\pi N} (F_1^{(v)}(k^2) + F_2^{(v)}(k^2)) \frac{1}{4} \left(\frac{1}{\nu_B-\nu} \pm \frac{1}{\nu_B+\nu}\right), \tag{3.20}$$

$$M_6^{(\pm)} = g_{\pi N} F_2^{(v)}(k^2) \frac{1}{4m_N} \left(\frac{1}{\nu_B-\nu} \pm \frac{1}{\nu_B+\nu}\right),$$

$$M_7^{(\pm)} = g_{\pi N} F_2^{(v)}(k^2) \frac{1}{8m_N} \left(\frac{1}{\nu_B-\nu} \mp \frac{1}{\nu_B+\nu}\right),$$

The amplitudes (0) follow from the (+) just by inserting the isoscalar form factors.

For the one-pion exchange diagram the only nonvanishing contributions are

$$M_3^{(-)} = -\frac{2g_{\pi N} F_\pi(k^2)}{m_\pi^2 - t} \quad ; \quad M_4^{(-)} = \frac{g_{\pi N} F_\pi(k^2)}{m_\pi^2 - t} \quad . \tag{3.21}$$

Although we shall often refer to the above expressions as the Born approximation, it is obvious that they are not simply connected with perturbation theory, owing to the presence of the form factors.

After these preliminaries, we write the dispersion relations, embodying the analyticity properties of the scalar amplitudes in the variable ν, at fixed t, k^2. Their prototype is

$$M_i^{(\pm)}(\nu,t,k^2) \doteq M_i^{(\pm)}|_{\text{Nucleon}} + \frac{1}{\pi} \int_{\nu_0}^{\infty} \text{Im } M_i^{(\pm)}(\nu',t,k^2) \left(\frac{1}{\nu'-\nu-i\varepsilon} \pm \eta_i \frac{1}{\nu'+\nu}\right). \tag{3.22}$$

In (3.22), ν_0 is the physical threshold for the reaction, $s_0 = (m_N + m_\pi)^2$, i.e., $\nu_0 = \nu_B + m_\pi m_N + m_\pi^2/2$, and the negative ν contribution has been reduced, via crossing, to the $\nu > 0$ range. Furthermore, the behaviour around the integration singularity ($\nu > \nu_0$) has been specified by the $i\varepsilon$ instruction.

The meaning of the symbol $\doteq$ is "apart from subtraction constants". The presence of such constants is related first of all to the need of guaranteeing the convergence of the dispersive integrals and depends therefore on the asymptotic behaviour of the *imaginary parts*. Other subtractions many be introduced by the asymptotic behaviour of the *real parts;* the pion pole term in (3.21) is a good example of a subtraction constant of this type (not affecting, of course, the imaginary parts). According to the point of view outlined in the Introduction we shall adopt the model where the asymptotic behaviour of the complete amplitude is only determined by exchange of Regge trajectories in the t channel. Then only the constants dictated by convergence reasons appear, and terms corresponding to the exchange in the t channel of the pion and other particles are naturally generated by the tail of the ν integral.

We defer to Appendix D a more detailed description of the asymptotic behaviour of the electroproduction amplitudes. It is not hard to see, combining those results with the present phenomenological indication that at small spacelike t the leading trajectories do not exceed unity ($\alpha(t) < 1$), that no subtractions are, in principle, required in the framework of the simple Regge pole model.

D) The above analysis of properties of the electroproduction amplitude has been devoted mainly to features which, even if of a kinematical nature, like the choice of the invariant functions, are actually deeply influenced by the dynamics of strong interactions. In this sense such an analysis is common to electroproduction and any other hadron process. We now turn to a discussion of the additional constraint, peculiar to electroproduction, which follows from the conservation of the electromagnetic current. As will be shown, some interesting consequences can be derived also in this case by combining that requirement with analyticity and other strong interaction properties of the amplitude.

Current conservation $\partial^\mu V_\mu^{em} = 0$ amounts to the requirement

$$k^\mu M_\mu = 0. \tag{3.23}$$

This immediately leads to the following constraint among the invariant amplitudes (3.16):

$$\begin{aligned} &\nu M_2 + q\cdot k\ M_3 + k^2\ M_4 = 0, \\ &M_1 + \nu M_6 + q\cdot k\ M_7 + k^2 M_8 = 0. \end{aligned} \tag{3.24}$$

We see then that while the elimination of M_1 is immediate, it is impossible to express one among the amplitudes M_2, M_3, M_4 in terms of the others without dividing by ν, $q\cdot k$ or k^2 and introducing therefore a spurious singularity. In reducing the expansion of the amplitude to a set of automatically gauge-invariant vectors, M_1 and M_3 are usually eliminated, but clearly the choice is not unique. Anyway, since a particularly useful parametrization, mainly in the calculation of the cross section, is obtained by introducing the six independent centre-of-mass amplitudes described in Appendix C, we do not insist on this point. Let us rather elaborate on the effective information one can extract from (3.24) and from the independent analyticity of all invariant amplitudes.

We consider here the case of a real photon $k^2 = 0$, while electroproduction is discussed in Appendix D. For $k^2 = 0$, (3.24) reads

$$\nu M_2 + q\cdot k\, M_3 = 0,$$
$$M_1 + \nu M_6 + q\cdot k\, M_7 = 0. \tag{3.25}$$

As already mentioned, elimination of M_2 or M_3 through the first relation would introduce unwanted singularities (even if harmless from a strictly practical point of view, since $\nu = 0$ and $q\cdot k = 0$, i.e., $s = \bar{s} = m_N^2$, $t = m_\pi^2$, lie outside the physical region). Let us thus investigate the behaviour as $\nu \to 0$, $q\cdot k \to 0$. The nontrivial character of these relations derives from the fact that the amplitudes involved are singular at these points. In particular, using the explicit polar expressions, one finds for the isospin antisymmetric[7] amplitudes

$$\lim_{\nu,\nu_B \to 0} \nu M_2^{(-)} = -g_{\pi N}\, F_1^{(v)}(0) \lim_{\nu,\nu_B \to 0} \frac{\nu^2}{\nu_B^2 - \nu^2},$$
$$\lim_{\nu,\nu_B \to 0} q\cdot k\, M_3^{(-)} = -g_{\pi N}\, F_\pi(0) + g_{\pi N}\, F_1^{(v)}(0) \lim_{\nu,\nu_B \to 0} \frac{\nu_B^2}{\nu_B^2 - \nu^2}. \tag{3.26}$$

Comparing, one has, from the first of (3.25)

$$F_1^{(v)}(0) = F_\pi(0). \tag{3.27}$$

which, introducing a common unit of charge, becomes

$$e_p - e_n = e_{\pi^+}.$$

Thus, as is obvious, current conservation guarantees the exact charge balance for the reaction /63/. Let us also notice that the condition (3.27) ensures that the Born terms are, in the case $k^2 = 0$, automatically gauge invariant for any value of ν, t. It is similarly fruitful to examine the second relation, which offers the possibility of expressing the amplitude M_1 in terms of the polar residues. A simple calculation shows that at the *unphysical* point $\nu = 0$, $t = m_\pi^2$,

[7] For the (+, o) amplitudes the requirement is trivial by crossing.

$$M_1^{(-)}\ (\nu=0,\ t=m_\pi^2) = g_{\pi N}\,\frac{F_2^{(v)}(0)}{2m_N},$$
$$M_1^{(+,0)}\ (\nu=0,\ t=m_\pi^2) = 0. \tag{3.28}$$

In the ideal case of a vanishing pion mass (to which we shall become accustomed soon), the point $\nu = 0$, $t = 0$ corresponds to the threshold of photoproduction and (3.28) represents a low-energy theorem, the Kroll-Ruderman theorem /64/. Indeed in the forward direction ($k\cdot\varepsilon = q\cdot\varepsilon = 0$) and in the gauge $p\cdot\varepsilon = 0$, the full amplitude takes the form

$$M\cdot\varepsilon = \bar{u}\gamma_5\ \gamma\cdot\varepsilon\ u\ (M_1 + \frac{\nu}{m_N}\,M_5).$$

Using (3.25) plus the value of νM_5 as $\nu\to 0$ [use polar expressions (3.20)], one finally obtains the well-known expression of the Kroll-Ruderman theorem

$$M^{(-)}\cdot\varepsilon = -\frac{g_{\pi N}}{2m_N}\ \bar{u}\ \gamma_5\ \gamma\cdot\varepsilon\ u = \frac{g_{\pi N}}{2m_N}\ \underline{\sigma}\cdot\underline{\varepsilon},\quad M^{(+,0)}\cdot\varepsilon = 0. \tag{3.29}$$

3.3 Dynamical Models

A detailed discussion on the different theoretical approaches leading to estimates of the full low-energy electroproduction amplitude lies outside the range of the present review. In this section we shall limit outselves to outlining the physical ideas and the main results.

Most work on the subject relies in one way or another on the relativistic dispersion relations described in the previous section. One is thus led to separate the full amplitude into a "Born term", given by the contribution of the nucleon pole plus possible subtractions, and the "dispersion correction", whose evaluation is the object of the different dynamical approaches.

A common assumption which is being made is that the dispersion integral is dominated by low-energy ($\lesssim$ 1 GeV) contributions. The following features are thus present:

1. In this region, single pion electroproduction is still the dominant process and we are in a situation in which (according to the Fermi-Watson theorem) each multipole has a phase equal to the final state pion-nucleon phase shift.
2. As a consequence the most important contribution to each dispersion integral comes from the production of resonant states, in particular from the celebrated $\Delta(3/2,\ 3/2)$ isobar which plays a fundamental role in low-energy pion physics. Its properties have been reviewed in Section 2.1.4.

The different investigations on low-energy electroproduction differ on how specifically each theoretical ingredient is being used and do therefore exhibit a different predictive power.

The most direct and elementary approach is to exploit directly the dominance of the different resonant states and to construct an isobaric model in which one introduces the diagrams due to the production of the different isobaric states /66/. Those diagrams are given in Fig. 1.3 d, where the isobaric propagator is the well-known Rarita-Schwinger spin-3/2 propagator. The ($\gamma N\Delta$) vertex contains the three transition form factors, $G_M^{(1)}(k^2)$, $G_E^{(1)}(k^2)$, $G_C^{(1)}(k^2)$, introduced in Section 2.1.4, (2.22), and the ($\gamma N\Delta$) vertex introduces the transition coupling constant g^* given in Section 2.1.4. This can lead in principle to a phenomenological determination of the electromagnetic transition isobaric form factors. A more refined but somewhat more cumbersome version of the same idea is to use the isobaric model in order to evaluate the imaginary part of the different invariant amplitudes. This avoids some of the ambiguities related to the propagator of an off-mass-shell isobar and can allow one to obtain some phenomenological relations between nucleon and isobar form factor through the use of the dispersion sum rules (see, e.g., Section 3.5.F). A large amount of theoretical work has been done by applying to electroproduction the programme based on the combination of analyticity and unitarity /61, 66/.

We have already seen that (in the elastic approximation) each electroproduction multipole, corresponding to a well-defined πN final state, must have a phase which is equal to the corresponding pion-nucleon phase shift. So the effective application of this programme requires the use of a multipole decomposition of the electroproduction amplitude. One must thus go through the painful operation of translating the simple dispersion relations for the invariant amplitudes of Section 3.2 into equivalent relations for the different multipoles.

If one denotes a multipole amplitude by $F_l(\omega)$, one is led to a dispersion relation of the form

$$F_l(\omega) = F_l^{B.}(\omega) + \frac{1}{\pi}\int \frac{\mathrm{Im}\{F_l(\omega')\}}{\omega'-\omega-i\varepsilon}\, d\omega' + \delta F_l, \tag{3.30}$$

where $F_l^{B.}(\omega)$ is the known contribution due to the nucleon pole and to possible subtraction constants, δF_l is the contribution due to the crossed term and to inelastic contribution, and, in the main dispersion integral, $\mathrm{Im}\{F_l\}$ is related to F_l through the celebrated Fermi-Watson theorem,

$$\mathrm{Im}\{F_l(\omega)\} = e^{-i\delta_l}\sin\delta_l\, F_l(\omega), \tag{3.31}$$

i.e.,

$$F_l(\omega) = e^{i\delta_l} R_l(\omega), \tag{3.32}$$

where $\delta_l(\omega)$ is the l-th. pion-nucleon phase shift and $R_l(\omega)$ is a real quantity.

If one disregards (in first approximation) δF_l, the combination of (3.30) and (3.31) can be exactly solved by means of the Muskhelishvili-Omnes formula.

A detailed discussion of this procedure can be found elsewhere /61/; here we wish to outline the following points:

a) From (3.31) one sees that $\mathrm{Im}\{F_l\}$ is large only if the corresponding scattering phase shift is large. We shall thus get a large correction to the Born term only for the multipoles leading to the resonant 3/2, 3/2 states. This property is of course directly exhibited by any isobaric calculations.

b) Both (3.30) and (3.31) are linear in $F_l(\omega)$. This means that the solution of such equations is linear in $F_l^B(\omega)$.

In other words, the only large multipoles are those corresponding to the large Born approximation and thus contributes effectively to the Δ production. Again this magnetic dipole dominance is empirically checked in the framework of isobaric models.

The main conclusion of this short discussion is that one expects that the physical electroproduction amplitude may differ strongly from the Born approximation mainly owing to the excitations of the Δ isobar. A careful discussion of the relative weight of the various contributions (Born terms and continuum integrals) to the cross sections in the regions of interest (threshold, first resonance region) can be found in /67/, whose results will be largely used in the following.

The nature of the physical process is thus quite simple and clear. On the other hand, as soon as one wishes to perform an accurate analysis of all the features of the electroproduction process, the technical details become so cumbersome that a complete description is impossible within the size of the present review.

3.4 The Role of the Pion

A. *Dynamical Aspects.* If we look at the table of hadronic particles, we see that the mass of the pion is abnormally small. The significant parameter m_π^2 is an order of magnitude smaller than the corresponding square masses of other nonstrange particles.

Independently of whether such a small mass has a deep physical origin or is a purely dynamical accident, it is clear that it gives to the pion a particularly important role in the analysis of many physical processes.

It is well known that the outer part of the nucleon cloud is completely dominated by the one-pion contribution. In the dispersion theory language, this means that it is possible, or at least conceivable, to isolate, at small momentum transfer t, the single-pion exchange pole in many processes /68/, illustrated in Fig. 3.1.

This situation came to a very exciting point when it became clear that weak currents have an axial component A_μ. Particular interest was devoted to a study of the properties of the divergence of the axial vector current,

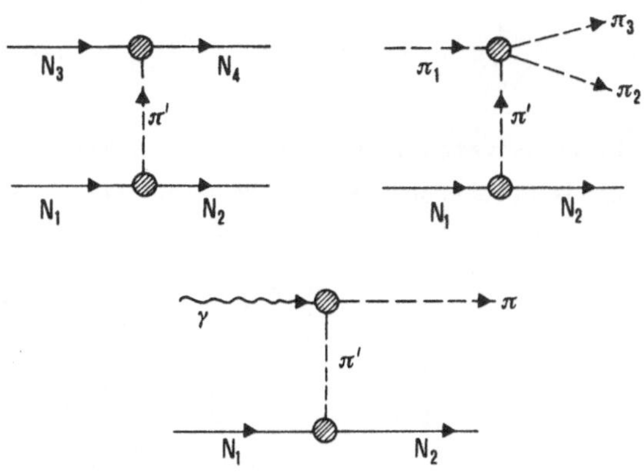

Fig. 3.1. Pion exchange contribution to a) nucleon scattering, b) pion production by pions, c) pion production by photons

$$\bar{D} = \partial^\mu A_\mu , \tag{3.33}$$

whose matrix elements between hadron states were found to be dominated by the pion pole /69/. The pion pole contribution involves, on the other hand, in an essential manner the matrix element of $\bar{D}$ between the vacuum and the one-pion state. Using the definition of the pion decay constant

$$\langle 0|A_\mu|\pi(q)\rangle = i f_\pi q_\mu , \tag{3.34}$$

this turns out to be

$$\langle 0|\bar{D}|\pi(q)\rangle = m_\pi^2 f_\pi . \tag{3.35}$$

The dynamical fact of the pion dominance of the hadronic matrix elements of $\bar{D}$[8] can be given a symbolic and more general form, by introducing the so-called PCAC (partial conservation axial current) relation /70/

$$\bar{D} = m_\pi^2 \, f_\pi \, \phi_\pi , \tag{3.36}$$

where ϕ_π represents an interpolating pion field ($\langle 0|\phi_\pi|\pi\rangle = 1$). The meaning of (3.36) is to express from the beginning the fact that via pole dominance and (3.35) *all* matrix elements of $\bar{D}$ are proportional to m_π^2. The first fundamental application of PDDAC (or PCAC) was obtained by considering the matrix element of $\bar{D}$ between nucleon states

[8] Sometimes called P(ion) D(ominance) D(ivergence) A(xial) C(urrent).

$$\langle N_2|\bar{D}|N_1\rangle = \frac{m_\pi^2 f_\pi}{m_\pi^2 - q^2} \, 2g_{\pi N} \, i\bar{u}_2 \gamma_5 u_1 . \qquad (3.37)$$

The graphical interpretation of (3.37) is illustrated in Fig. 3.2. Eq. (3.37), when taken at $(p_2-p_1)^2 \equiv q^2 = 0$, leads to the famous Goldberger-Treiman relation /72/

$$g_{\pi N} \approx \frac{MG_A(0)}{f_\pi} , \qquad (3.38)$$

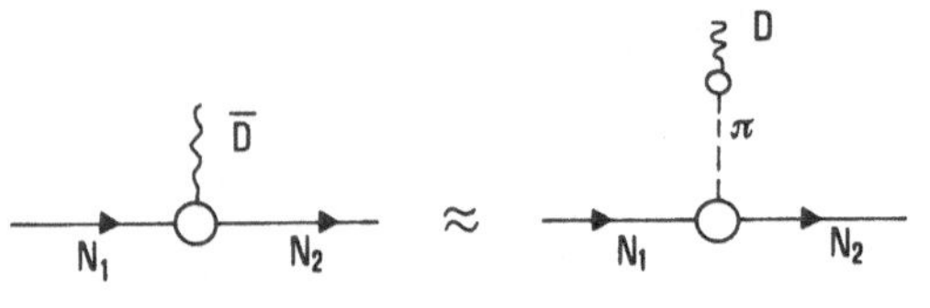

Fig. 3.2. Pion dominance for the axial divergence form factor

which allows us to evaluate the pion-nucleon coupling constant, a strong interaction quantity, in terms of the weak interaction parameters f_π, $G_A(0)$. The agreement between the theoretical prediction and the experimental value is quite reasonable, within 7 % [with $(\sqrt{2}f_\pi)_{exp.} = (0.9442 \pm 0.0008)\ m_\pi$ /20/ one finds $(g_{\pi N})_{G.T.} \approx 12.7$, while $(g_{\pi N})_{exp} \approx 13.5$]. It is important to notice that the representation (3.37) for $\langle N_2|\bar{D}|N_1\rangle$ should be adequate for small q^2, when the small denominator ($\approx m_\pi^2$) compensates the small numerator ($\approx m_\pi^2$). In particular at $q^2 = 0$ the pion mass factor disappears, while this does not occur for higher states.

A very important application of the same point of view leads to the so-called Adler consistency relation /72/

$$iq^\mu \langle N_2\pi(k)|A_\mu|N_1\rangle = \langle N_2\pi(k)|\bar{D}|N_1\rangle \approx \frac{f_\pi m_\pi^2}{m_\pi^2 - q^2} T_{\pi N}, \qquad (3.39)$$

where $T_{\pi N}$ is the pion-nucleon scattering amplitude. Eq. (3.39) tells us that in the limit $q^\mu \to 0$, the πN amplitude tends to zero[9]. If for the moment we disregard the fact that such a limit can only be reached with zero mass pions, we find the important result that the threshold pion-nucleon amplitude vanishes. Introducing the

[9] Such a statement actually requires some care, since other singular terms, like the nucleon Born diagram, can in principle contribute. This is a consequence of considering a scattering process where more kinematical variables depend on q_μ. More refined considerations show that (3.40) does indeed hold for zero-mass pions.

s-wave scattering lengths a_1, a_3 for isospin 1/2, 3/2 states, for the symmetric combination one gets the result

$$a_1 + 2a_3 \approx 0, \qquad (3.40)$$

again in agreement with experiment [$(m_\pi a_1)_{exp} = 0.171 \pm 0.005$, $(m_\pi a_3)_{exp} = -0.088 \pm 0.004$]. Formula (3.40) partially explains a long-standing paradox (known to old-timers as pair suppression): why are threshold s-waves so small and why does the isospin antisymmetric combination dominate? To use (3.39) in a more complete manner, one must apply a similar trick to the second pion.

The final results turns out to be

$$\lim_{k \to 0} \langle N_2 \pi^\beta(k)|A_\mu^\alpha|N_1\rangle = -i\varepsilon_{\alpha\beta\gamma}\langle N_2|V_\mu^\gamma|N_1\rangle, \qquad (3.41)$$

where use has been made of the GELL-MANN /73/ recipe for the equal-time commutator between two axial current densities

$$[A_o^\beta(\bar{x}),\ A_o^\alpha(0)] = i\varepsilon_{\beta\alpha\gamma}\ V_\mu^\gamma(0)\delta(\underline{x}). \qquad (3.42)$$

Eq. (3.41) finally inserted in (3.39) reproduces the well-known and successful TOMOZAWA-WEINBERG prediction for the isospin odd combination of pion-nucleon scattering lengths /74/

$$a_1 - a_3 \approx \frac{3}{8\pi}\ \frac{m_\pi}{f_\pi^2} \approx 0.25\ m_\pi^{-1}. \qquad (3.43)$$

Of course, similar reasonings, to be treated in more detail in the following, lead to equally interesting formulae for electro- and photoproduction of pions.

The beauty and simplicity of the previous formulae, together with the success of the strategy based on the exploitation of the small pion mass, strongly suggest that those results have a deep, more fundamental root.

On the other hand, we shall be faced with the practical question of obtaining a reasonable estimate of the main corrections arising from the presence of a small but not vanishing pion mass. Both problems will be dealt with in the next sections.

B. *The Mechanism for Axial Current Conservation.* In the previous section we have seen some examples of the intimate connection between axial currents and soft pions. Exploiting the general properties of the axial current and pion pole dominance, we have found a few remarkable low-energy theorems for soft-pion interaction. Similar theorems exist for electroproduction or neutrino production of pions, as we shall see later, and in general for a large class of weak phenomena where pions appear among the decay products. Before proceeding further it is, however, important to gain some understanding of the deep origin of those pion properties.

Let us recall some of the well-known properties of the vector and axial vector currents we use to describe the weak and electromagnetic interactions between hadrons and leptons. These currents are assumed to belong to a larger set of currents V_μ^α, A_μ^α which, according to the fundamental GELL-MANN-CABIBBO /75/ hypothesis, transform as octets under SU(3), with $\alpha = 1...8$. If, furthermore, one considers the associated charges

$$Q_\alpha = \int d^3x\ V_0^\alpha(x), \quad \bar{Q}_\alpha = \int d^3x\ A_0^\alpha(x), \tag{3.44}$$

these objects are assumed to be the generators of the SU(3)xSU(3) algebra and to obey the following set of equal-time commutation relations [$f_{\alpha\beta\gamma}$ are the SU(3) structure constants]:

$$\begin{aligned} [Q_\alpha, Q_\beta] &= i f_{\alpha\beta\gamma}\ Q_\gamma, \quad \alpha,\beta,\gamma = 1...8, \\ [Q_\alpha, \bar{Q}_\beta] &= i f_{\alpha\beta\gamma}\ \bar{Q}_\gamma, \\ [\bar{Q}_\alpha, \bar{Q}_\beta] &= i f_{\alpha\beta\gamma}\ Q_\gamma. \end{aligned} \tag{3.45}$$

One has, similarly,

$$\begin{aligned} [Q_\alpha, V_\mu^\beta(\bar{x})] &= i f_{\alpha\beta\gamma}\ V_\mu^\gamma(\bar{x}), \\ [Q_\alpha, A_\mu^\beta(\bar{x})] &= i f_{\alpha\beta\gamma}\ A_\mu^\gamma(\bar{x}), \\ [\bar{Q}_\alpha, A_\mu^\beta(\bar{x})] &= i f_{\alpha\beta\gamma}\ V_\mu^\gamma(\bar{x}) \end{aligned} \tag{3.46}$$

This identification represents a nontrivial connection between the symmetry properties of the hadron world and the quantities used to describe the lowest order weak and electromagnetic interaction of hadrons. Such an assumption generalizes the well-known C.V.C. /76/ (conserved vector current) properties of the weak nonstrange vector current, which must be identified with the isospin current.

All these currents exhibit more or less rigorous conservation properties and, as a consequence, the corresponding charges are approximate constants of the motion, reflecting in turn underlying symmetry properties of hadrons.

To simplify the discussion, we shall work in the idealized situation of exact current conservation. In this case the charges commute with the four-momentum operator P_μ

$$[Q_\alpha, P_\mu] = 0. \tag{3.47}$$

As a consequence the only nonvanishing matrix elements of the charges Q_α are those between states of equal four-momentum. This means that in the case *where zero-mass particles are absent,* the only one-particle matrix elements of the Q_α are between particles of the same mass. We thus see that on the basis of the commutation relations (3.45) one can infer the existence of supermultiplets of equal mass particles with different internal quantum numbers, while (3.46) show that different matrix elements are related by the Wigner-Eckart type relations.

For example, in the case of vector charges, the combination of (3.45) and (3.46) forces the existence of exact SU(3) multiplets and reproduces the familiar SU(3) relations between vertices and coupling constants.

The presence of zero-mass particles can change the whole picture. Indeed, the possibility exists now of nonvanishing matrix elements between the vacuum state and the one-particle state at zero four-momentum. This possibility is extremely important for us. We believe that axial charges, at least at the nonstrange level, are, with good approximation, constants of motion. On the other hand, there is no firm evidence of the existence of approximate supermultiplets of particles of opposite parity (for instance, the nearest $1/2^-$ nucleon partner is at 1570 GeV). The most natural way out of such a situation is to rely on the presence of a pseudoscalar particle of small mass, *the pion.* We shall thus be led to the conclusion that the pion mass vanishes in the limit of exact axial current conservation /77/ $\partial^\mu A^\alpha_\mu = 0$, $\alpha = 1, 2, 3$.

Let us consider the situation in more detail[10]. If the axial charges are exactly conserved and no parity supermultiplet exists the only nonvanishing matrix element of the charge is

$$<0|\bar{Q}_\alpha|\pi_\beta> = i\delta_{\alpha\beta}\ f_\pi\ E_\pi\ (2\pi)^3\delta(\underline{\pi}). \tag{3.48}$$

Let us then start with the matrix element of a commutator

$$<B|[\bar{Q},\ F]|A> = C^F_{BA} \tag{3.49}$$

[where F is any operator with definite transformation properties under SU(2)xSU(2)] and compute it by inserting a complete set of intermediate states. The only contributions turn out to be those containing the matrix element (3.48) both in the direct and in the crossed term, and we obtain

[10] We disregard the difficulty arising from the fact that in the exact limit $m_\pi = 0$, the axial charges does not properly exist. In practice, everything works fine if we start with a small value of m_π and smoothly reach the limit $m_\pi = 0$.

$$<B|[\bar{Q}, F]|A> \approx <0|\bar{Q}|\pi><\pi B|F|A> + <B|F|A\pi><\pi|\bar{Q}|0> = C^F_{BA}. \tag{3.50}$$

One thus sees that the knowledge of the commutator (3.49) provides a direct information on the matrix element where a soft pion is either emitted or absorbed. In the case of electroproduction of soft pions, the relevant commutator is $[\bar{Q}_\alpha, V^{em}_\mu] = i\varepsilon_{\alpha 3\beta}A^\beta_\mu$ which, when sandwiched between nucleon states, will lead to a direct connection between electroproduction of a soft pion and axial nucleon form factors.

In the realistic situation of non-massless pions, i.e., of not perfectly conserved axial current, the nucleon matrix element of $[\bar{Q}_\alpha, V^{em}_\mu]$ can still be used to obtain an improved form of the soft-pion theorems. Selection of the one-pion contribution will reproduce the soft-pion results, while the other contributions will give rise to possible corrections (vanishing as $m_\pi \to 0$) to those low-energy theorems. We turn now to a more detailed analysis of this matter.

3.5 Low-Energy Electroproduction and Current Commutators

We devote this section to a detailed application of the above ideas to the electroproduction process. In particular it will be shown how to derive, from the equal-time commutator between the axial charge and the electromagnetic current, a simple representation for pion electroproduction amplitudes in the threshold region. For the sake of generality (and also for physical reasons!) we shall consider from the beginning the case of a partially conserved charge, i.e., of pions of physical, non-vanishing mass. The exact low-energy theorems, a consequence of chiral invariance, will then follow in the limit $m_\pi \to 0$; however, the procedure is quite general and provides a recipe to evaluate the effects due to the finite pion mass, i.e., the "corrections" arising from chiral breaking.

A) To derive our result we shall resort to the simple technique, which is based on direct saturation of the equal-time commutator of interest taken between one-nucleon states. Other, no doubt more elegant, techniques exist but the final outcome is essentially equivalent. Let us therefore start from the commutator identity

$$M_\mu = <N_2|[\bar{Q}_\alpha, V^{em}_\mu(0)]|N_1> = i\varepsilon_{\alpha 3\gamma} <N_2|A^\gamma_\mu|N_1> \tag{3.51}$$

and use completeness

$$M_\mu = \Sigma_n <V_2|\bar{Q}_\alpha|n><n|V^{em}_\mu|N_1> - \text{c.t.} = i\varepsilon_{\alpha 3\gamma}<N_2|A^\gamma_\mu|N_1>. \tag{3.52}$$

The general structure of the completeness sum is not as simple as one would expect, for instance, from the analogy with the nonrelativistic sum rules of nuclear or atomic physics. More complicated types of contributions can now be present and the simple reason is the fact that the number of particles is not conserved in a relativistic theory, owing to the phenomenon of annihilation and creation of quanta.

Indeed, the completeness sum contains the part one would have naively guessed, i.e., the so-called connected contributions[11]

$$M_\mu^I = \Sigma_\alpha <N_2|\bar{Q}_\alpha|\alpha>_c <\alpha|V_\mu^{em}|N_1>_c - c.t. \tag{3.53}$$

whose form is illustrated in Fig. 3.3 and where the lowest intermediate state is, of course, the nucleon.

But besides that, there are more complicated contributions and among them the disconnected diagrams shown in Fig. 3.4,

$$M_\mu^{II} = \Sigma_{\gamma_1} <N_2|\bar{Q}_\alpha|N_1\gamma_1>_c <\gamma_1|V_\mu^{em}|0> + \Sigma_{\gamma_2} <0|\bar{Q}_\alpha|\gamma_2><\gamma_2 N_2|V_\mu^{em}|N_1>_c - c.t. \tag{3.54}$$

The first contribution corresponds to the creation from the vacuum of a hadron system with the same quantum numbers as the electromagnetic current (the ρ,ω,ϕ ...) followed by the process $\alpha_1+N_1 \rightarrow \bar{Q}+N_2$ (just vector dominance). Similarly, the second term, which is the one of interest to us, describes the creation, among other states, of a *physical pion at rest* from the vacuum and explicitly introduces in the game the physical amplitudes

$$T_\mu^\alpha = i<N_2\pi^\alpha|V_\mu^{em}|N_1>_c, \qquad \bar{T}_\mu^\alpha = i<N_2|V_\mu^{em}|N_1\pi^\alpha>_c. \tag{3.55}$$

Because of the small pion mass, this last contribution will be the dominant one, actually the only one surviving as $m_\pi \rightarrow 0$. (For instance, the next multipion states $|\gamma_2> = |3\pi>$... should be depressed by phase space reasons.)

M_μ^I, M_μ^{II} or (3.53) and (3.54) do not exhaust the terms present in the completeness sum. There are also the so-calle Z diagrams, but since they do not play an essential role in the following considerations, we shall not discuss them.

B) The explicit selection of the disconnected pion contributions leads to an improved version of the pion low-energy theorem, in a sum rule form,

$$\frac{1}{2f\pi}\left[T_\mu^\alpha(\underline{q}_\pi = \underline{0}) + \bar{T}_\mu^\alpha(\underline{q}_\pi = \underline{0})\right] = i\varepsilon_{\alpha 3\gamma}<N_2|A_\mu^\gamma|N_1> - \Sigma_n <N_2|\bar{Q}_\alpha|n><n|V_\mu^{em}|N_1> - c.t. \tag{3.56}$$

---

[11] We recall that, given the matrix element $T_A = <p_2,\beta|A|p_1>$, where p_1, p_2 are momenta belonging to identical particles, its connects part is defined as follows:

$T_A^c \equiv <p_2,\beta|A|p_1>_c = T_A - <p_2|p_1><\beta|A|0>$.

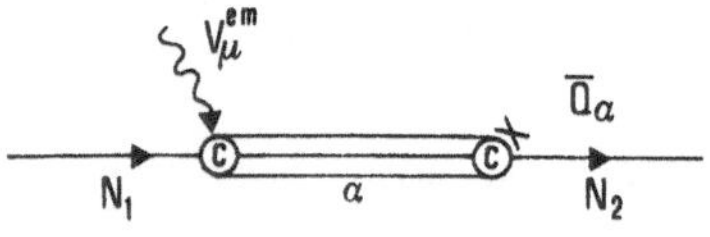

Fig. 3.3. Connected contributions to the completeness sum

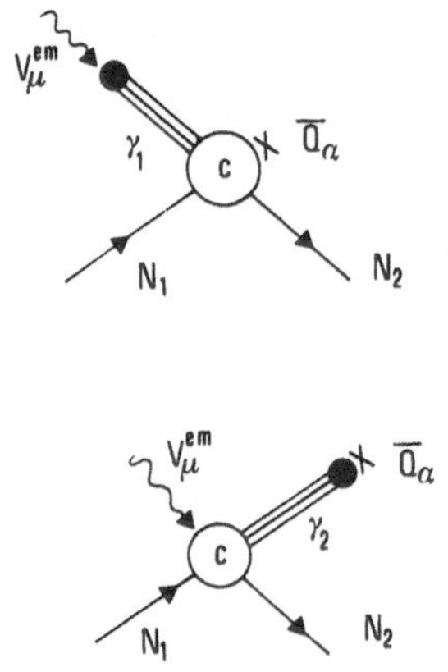

Fig. 3.4. Disconnected contributions

The pion amplitudes are calculated for physical pions at rest, outgoing, and ingoing, which in terms of the familiar invariant variables corresponds to

$$\nu = \pm\, m_\pi P_0, \qquad t = \Delta^2,$$
$$(k^2)_\pm = (\Delta \pm q)^2 = m_\pi^2 + t \pm 2m_\pi \Delta_0. \tag{3.57}$$

This shows that the precise point where T_μ^α and $\bar{T}_\mu^\alpha$ are evaluated still depends on the configuration of the external nucleon. Such a *"frame dependence"* represents a long-standing and well-known feature of the sum rule approach to the equal-time commutators of a relativistic theory. The reason for that is not hard to understand and lies in the fact that commutators at equal times are not Lorentz invariant (unless conserved time-independent charges are involved). As a consequence, the evaluation of the completeness sum, in particular the relative weight of the various contributions, depends strongly on the external nucleon configuration. Different choices can thus be performed according to the different aspects one wants to emphasize.

Since, in this book, we are mainly interested in having a representation for the pion amplitudes, as simple as possible, we suggest to work in the *Breit frame* $\underline{P} = 0$ *of fairly slow* nucleons /78/. This choice gives $|\underline{p}_1| = |\underline{p}_2|$, i.e., $\Delta_0 = 0$, $t = -4\underline{p}^2$, so that

$$k_+^2 = k_-^2 = t + m_\pi^2.$$

Then T^α_μ and $\bar{T}^\alpha_\mu$ differ only by $\nu \to -\nu$ and are related by crossing. It is also important to note that small values of $t = -4\underline{p}^2$ offer the advantage of an approximate selection rule for the connected contributions, due to parity and angular momentum conservation. This follows from the nature of the axial charge: given a matrix element $\langle p|\bar{Q}_\alpha|n\rangle$, for which $\underline{p}_n = \underline{p}$, only s-wave excitation is allowed in the limit $\underline{p} \to 0$. Thus only states with the same spin as the target and opposite parity contribute,[12] in particular $\frac{1}{2}^-$ baryon states for the direct contribution. Since in our case $\underline{p}_1$ and $\underline{p}_2$ cannot be taken simultaneously as zero[13], a strict selection rule does not hold: however, even if p- [the Δ(3,3) for example] and higher waves are in principle allowed, their contribution will be strongly depressed working with small $|t| = 4\underline{p}^2$.

Concluding, our prediction will concern electroproduction at the "Breit threshold" (pion at rest produced in the Breit frame of the external nucleon) corresponding to the following values of the invariant variables:

$$(\nu)_{B.th.} = m_\pi P_0 = m_\pi M \left(1 - \frac{k^2 - m_\pi^2}{4M^2}\right)^{1/2}, \qquad (t)_{B.th.} = k^2 - m_\pi^2. \tag{3.58}$$

In particular, the photoproduction limit is reached with

$$k^2 \to 0, \quad t \to -m_\pi^2.$$

It is easy to check that the Breit threshold configuration corresponds, for instance in C.M.S., to the pion emitted along $\underline{p}_1$, $\underline{q} = -\underline{p}_2$, i.e., $\theta_{CM} = \pi$, with energy $\omega_{CM} \approx m_\pi (1 - t/8M^2)$.

Therefore, as long as we consider a not too large $|t|$ ($|t| \lesssim 20\, m_\pi^2$), we can reasonably expect our predictions to hold without significant changes at the actual threshold[14].

C) We now come to the point concerning the behaviour of the various contributions to the sum rule in the limit $m_\pi \to 0$. We take advantage of the relation

$$d\bar{Q}_\alpha/dt = \int d^3x\, \bar{D}^\alpha(x) \tag{3.59}$$

to derive the simple identity

---

[12] A remarkable exception occurs if the target "nucleon" becomes massless (the neutrino!).

[13] The forward configuration $\underline{p}_1 = \underline{p}_2$, $t = 0$, is not allowed by the electroproduction kinematics.

[14] For instance, for $|t| \lesssim 20\, m_\pi^2$, $\omega_{CM} \lesssim 145$ MeV, $|\underline{q}|_{CM} \lesssim 40$ MeV.

$$\langle b|\bar{Q}_\alpha|a\rangle = \frac{\langle b|\dot{\bar{Q}}_\alpha|a\rangle}{i(E_b-E_a)} = (2\pi)^3\delta(\underline{p}_b-\underline{p}_a)\,\frac{\langle b|\bar{D}_\alpha|a\rangle}{i(E_b-E_a)}\ . \tag{3.60}$$

One can therefore conclude that matrix elements, for which the energy difference E_b-E_a is not vanishing, are of the order of the symmetry breaking and do not contribute in the symmetry limit $\bar{D}\to 0$. In our case this means that, with the exception of the nucleon, all contributions on the r.h.s. of (3.59) are $\propto\langle b|\bar{D}|a\rangle$, i.e., of order m_π^2, and will therefore play the role of corrections to the chiral symmetry result.

Time is ripe finally to be more explicit about the spin structure of the Breit threshold amplitude and the form of the various contributions. We shall adopt for T_μ^α the gauge invariant decomposition[15]

$$T_\mu^\alpha = u(p_2)\gamma_5\left[(\gamma_\mu + \frac{2m_N}{t}\Delta_\mu)\,T_1^\alpha + 2m_N\,(\Delta_\mu/t - P_\mu/\nu)\,T_2^\alpha\right]u(p_1). \tag{3.61}$$

[all quantities are evaluated at the values (3.58) of the scalar variables].

It is then straightforward to obtain for these amplitudes the representations below, which follow from the explicit selection in (3.56) of the nucleon matrix elements

$$T_1^{(-)}\ (\text{B.th.}) = \frac{G_A(t)}{2f_\pi} + \frac{t}{8E^2f_\pi}\,G_A(0)G_M^v(t) + \delta T_1^{(-)}, \tag{3.62}$$

$$T_2^{(-)}\ (\text{B.th.}) = \frac{D(t)}{4Mf_\pi} + \delta T_2^{(-)}, \tag{3.63}$$

$$T_2^{(+,0)}\ (\text{B.th.}) = \frac{m_\pi}{4Ef_\pi}\,G_A(0)G_E^{(v,s)}(t) + \delta T_2^{(+,0)} \tag{3.64}$$

Eqs. (3.62), (3.63), and (3.64) exhaust the information we can get from the fundamental charge-current commutator (3.51).

The correction terms δT_i are of course vanishing as $m_\pi\to 0$, and one can easily ascertain that formally $\delta T_{1,2}^{(-)}\sim O(m_\pi^2)$, $\delta T_2^{(+,0)}\sim O(m_\pi^3)$. It is also interesting that these quantities can be given a compact form as dispersive integrals at fixed t and variable q_0, subtracted at $q_0 = m_\pi$, for instance,

[15] In the Breit frame
$\underline{T} = -(E/m_N)[\underline{\sigma}-\underline{n}(\underline{\sigma}\cdot\underline{n})]T_1 + \underline{n}(\underline{\sigma}\cdot\underline{n})T_2$, $\underline{n} = \underline{k}/|\underline{k}|$,
showing the transversal and longitudinal parts of the expansion.

$$\delta T_i^{(-)} = 2m_\pi^2 \int_{m_\pi}^{\infty} \frac{\rho_i^{(-)}(q_0)dq_0}{(q_0^2-m_\pi^2)\, q_0} , \tag{3.65}$$

with $\rho_i(q_0)$ a well-defined spectral density which can be explicitly evaluated.

We thus see that, from this point of view, the great progress due to current algebra and chiral symmetry consists of having provided for the relevant amplitudes the subtraction constants in terms of physically meaningful quantities such as nucleon form factors. The price to pay has been the complicated form of the correction integrals [q_0 variable means ν, $q^2 = \nu^2/P_0^2$, $k^2 = t+q^2$ variable and the completeness sum in $\rho_i(q_0)$ must include the contributions of the type listed in Section 3.5.A]. However, since these terms are rapidly convergent integrals, weighted by factors m_π^2 (and higher waves are also depressed), one expects that they can in any case provide a reasonable indication on the size of the corrections.

As a final remark one must mention that no determination of the amplitudes $T_1^{(+,0)}$ has been possible starting from (3.51). The knowledge of more complicated commutators is required to achieve the goal, in particular of the quantity $[\dot{Q}_\alpha, \underline{V}_{em}]$. Such a quantity is clearly beyond the framework of current algebra and can be evaluated by resorting to specific models for the hadron currents and divergences. The quark model seems to be the preferred one and, using it /79/, the outcome is the following expression of $T_1^{(+,0)}$ [16]

$$T_1^{(+,0)} = -\left(\frac{\bar{m}}{m_\pi}\right) \frac{2m_N}{Ef_\pi} G_T(t) + \delta T_1^{(+,0)}, \tag{3.66}$$

where $G_T(t)$ is a form factor related to the nucleon matrix element of a tensor current, reproducing the exchange of ρ,ω,ϕ ... in the t-channel, and $\bar{m} \sim O(m_\pi^2)$ a parameter introduced by the quark model relation for the axial divergence

$$\bar{D}_\alpha = \bar{m}\, i\, \bar{q}\, \gamma_5 \tau_\alpha q. \tag{3.67}$$

In the above expression, which is quite similar to that of purely lepton theories, "$\bar{m}$" plays the role of the bare quark-proton mass, manifesting itself in chiral breaking and other hadron current phenomena. Finally, $\delta T_1^{(+,0)} \sim O(m_\pi)$, showing that there is no clear-cut separation in this case between the "subtraction" term (which is itself or order m_π) and the continuum.

16 In addition, an independent representation for $T_2^{(+,0)}$ can be derived, where consistency with (3.64) leads to a sum rule, which could in principle allow a determination of the interesting quantity $\bar{m}$. Isospin symmetry has been used to ascribe a common mass $\bar{m}$ to the proton- and neutron-quark.

Some activity has recently been devoted to a quantitative determination of $\bar{m}$, both to gain an improved understanding, at the constituent level, of the symmetry hidden beyond the small pion mass and to have a hint about relativistic quark dynamics. The available theoretical determinations of $\bar{m}$ /80/ are not unanimous ($\bar{m} \lesssim$ 20 MeV and $\bar{m} \cong m_\pi$ have been suggested), and low-energy electroproduction could perhaps provide, at a later stage, additional information on this point.

D) Our formulae can be considered as the extrapolation to the physical region of the soft-pion theorems, which would be obtained putting $m_\pi = 0$, $\nu = 0$, at fixed $k^2 = t$, in (3.62), (3.63), and (3.64). The appropriate way to exploit these relations is to give them the form of sum rules by using fixed-t dispersion relations. This application, which leads to algebraic relations between weak and electromagnetic vertices of hadrons, will be considered later. We rather turn now to a short discussion on other formulations of the low-energy theorems as well as of the approaches used to perform the extrapolation from the point $\nu = 0$, $q^2 = 0$, $k^2 = t$ to the physical region (this means that threshold electroproduction experiments will be a test of the chiral symmetry plus current algebra formulae *and* of the method used to continue them).

We begin by quoting the pioneer works by NAMBU and LURIE and NAMBU and SHRAUNER /81/, who first derived low-energy theorems for the threshold electroproduction multipoles in the limit of massless pions. In particular they obtained the simple and beautiful formula, which started the interest in the connection between pion electroproduction and the axial vector form factor $G_A(t)$, namely

$$E_{0+}^{(-)}(m_\pi=0) = \sqrt{\frac{4m_N^2-k^2}{4m_N^2}}\;\frac{g_{\pi N}}{2m_N}\left[\frac{G_A(k^2)}{G_A(0)} + \frac{k^2}{4m_N^2-2k^2}\;G_M^V(k^2)\right] . \qquad (3.68)$$

The explicit form of their results actually looks a little different from the soft-pion limit of ours; the reason for this can be traced back to the different frame used to implement chiral (and gauge) invariance /82/ on the relevant amplitudes. Subsequently a more refined version has also been proposed /83/, with a prescription to take finite pion mass effects into account.

In general the methods of extrapolation are based either on dispersion relations or on phenomenological Feynmann diagram models. The previous formulae, (3.62) and (3.64), are indeed an example of extrapolation along a curved path in the ν-q^2 plane.

An alternative procedure is to represent the threshold amplitude by ν-dispersion relations at fixed masses and momentum transfer with the soft-pion limits playing the role of subtraction constants /84/. Of course one must also take care of the extrapolation in the pion mass, from $q^2 = 0$ to $q^2 = m_\pi^2$, and in the variable $\nu_B = -q\cdot k/2$, from $\nu_B = 0$ (i.e., $t = k^2$) to $\nu_B = -m_\pi^2/2$ for the Breit threshold.[17] The distinctive

---

17 Or to $\nu_B = -\frac{m_\pi}{4}\left(\frac{k^2+m_N^2+2m_N m_\pi}{m_N+m_\pi}\right)$ for the effective threshold.

feature of this approach is that the smoothness of the various extrapolations is in some way related to the asymptotic behaviour of the relevant amplitude for $\nu \to \infty$ at fixed t, to be described by the Regge pole model. By this argument the low-energy limit for $T_2^{(-)}$ is not used and the form factor D(t) is not introduced in the theoretical formulae.

Thus the q^2 and ν_B extrapolations are performed directly on the dispersive integrals. These involve the imaginary parts of the electroproduction multipoles in the low-energy region, while the high-energy tails are expressed through the residues and the trajectory functions of the pion and other Regge poles[18] (for a similar discussion see Appendix D). In so doing the pion electromagnetic form factor $F_\pi(k^2)$ appears.

The approach based on the use of polar Feynmann diagrams has its starting point in a Lagrangian which is gauge and chiral invariant. The last property requires, as is well known, a pseudovector pion-nucleon coupling $L_{PV} = f\bar{\psi}_N\gamma_\mu\gamma_5\psi_N\partial^\mu\phi_\pi$. The minimal coupling recipe $\partial_\mu \to \partial_\mu - iea_\mu$ then gives rise to a $(NN\pi\gamma)$ contact term, which is identified as the analogue of the axial nucleon vertex deriving, in our approach, from the equal-time commutator. The model has also been extended to the first resonance region by including a Δ-pole diagram to take into account M_{1+} rescattering /86/.

A comparison between the predictions offered by the various models is deferred to Section 5. We can remark, however, that one of the features distinguishing other methods from that discussed in the previous pages is the different way in which the t-channel pion pole is introduced in the longitudinal amplitude. In a dispersion or Feynmann-like approach its presence is uniquely accompanied by the pion electromagnetic form factor $F_\pi(k^2)$, while the weak form factor D(t), which according to (3.63) embodies the pion singularity of the commutator term, does not exhaust the pion pole structure of the full amplitudes. In other words, higher terms will also contain the pion pole, and the relevant residues could be considered as corresponding to the higher moments of $F_\pi(k^2)$.

E) Equations (3.62), (3.63), and (3.64) are a set of predictions of chiral symmetry extrapolated to the physical low-energy region, and we want to discuss here some of their quantitative aspects.

There is the fact that our formulae actually hold for small, nonvanishing $|q_{C.M.}|$ so that, strictly speaking, T_1 and T_2 contain, besides the threshold multipoles $E_{0+}(q_{C.M.} = 0)$ and $L_{0+}(q_{C.M.} = 0)$, all higher terms in a $q_{C.M.}$ expansion. However, since we shall limit our analysis to $q_{C.M.} \lesssim 40$ to 50 MeV (and also experimentally such a kinematical situation can hardly be discriminated from the effective threshold),

---

[18] These quantities are usually taken to be very slowly varying functions of q^2 and ν_B /85/.

it is reasonable to expect that our fomulae can be used practically at threshold.

This statement has been quantitatively verified by deriving a generalized current algebra representation at variable $\underline{q}_{Breit}$, i.e., by letting the pion move in the Breit frame /87/. The point $\underline{q}_{C.M.} = 0$ can then be reached choosing $\underline{q}_B^2 = (m_\pi/2m_N)\underline{q}_B\cdot\underline{\Delta} = (m_\pi^2/4m_N^2)(m_\pi^2-k^2)(1+m_\pi/m_N)^{-1}$, but a larger energy range can actually be covered. from threshold until the beginning of the first resonance. Correspondingly, one can obtain predictions also for higher multipoles $M_{1\pm}$, $L_{1\pm}$ .... The formulae become more complicated, however, and the connection with chiral symmetry less transparent. For this reason we preferred to exhibit only the simple Breit threshold formulae, although for numerical evaluation the more general results have been used.

Furthermore, one must mention a sharp difference as far as the theoretical predictions for charged and neutral pions are concerned, both in photo- and electroproduction. For charged pion production the presence of the amplitude $T_1^{(-)}$ introduces a leading term $\propto G_A(t)$, to which the (o) part and the $\delta T_1^{(-)}$ terms are added as corrections: although important for a correct interpretation of experiments, these do not exceed 20 to 30 %. In the case of neutral pions, on the contrary, the continuum contribution can become larger than the equal-time and nucleon terms, which makes the final predictions strongly dependent on the approximations used to evaluate those integrals. For these reasons the theory can only provide, for neutral pions, indicative estimates,[19] and the best we can say is that there is only a qualitative agreement between theory and experiment, both suggesting very small values for threshold π^0 production.

It is useful to discuss first photoproduction (which in our formulation corresponds to $t = -m_\pi^2$). We shall directly use the relation

$$\sigma \equiv \left(\frac{|\underline{k}|}{|\underline{q}|}\,\frac{d\sigma}{d\Omega}\right)^{th.}_{C.M.} = \frac{\alpha}{4\pi}\left(\frac{m_N}{m_N+m_\pi}\right)^2 |E_{0+}(th.)|^2 \approx \frac{\alpha}{4\pi}\,\frac{E^2}{W^2}\,|T_1(B.th.)|^2 \tag{3.69}$$

and in Table 3.2 we reproduce some theoretical predictions and experimental results for the threshold cross sections.

The calculation uses, for the relevant parameter, the value $G_A(0)/2f_\pi \approx 0.92\ m_\pi^{-1}$, and the t-dependence of the axial vector form factor has been parametrized through a dipole fit

$$G_A(t) = G_A(0)\ (1 - t/M_A^2)^{-2}, \tag{3.70}$$

[19] Incidentally, a similar comment can be applied to dispersive calculations of lowenery π^0 photoproduction, where the large Born term is known to be drastically reduced by the s,p-wave dispersive contributions /88/.

Table 3.2

M_A	$6\ m_\pi$	$7\ m_\pi$	$8\ m_\pi$	$9\ m_\pi$	K.R.	exp.
$\sigma(\pi^+)$ [μb/sr]	15.2	15.6	15.8	16	13.5	15.6 ± 0.5 /89/
$R = \frac{\sigma(\pi^-)}{\sigma(\pi^+)}$	1.35	1.35	1.34	1.34	1	1.265 ± 0.75 /90/ 1.35
$\sigma(\pi^0 p)$ [μb/sr]	0.02 to 0.04				0	0.07 ± 0.02 /91/

but the dependence on M_A is rather weak. As an indication we have also included in Table 3.2 (sixth column) the prediction of the Kroll-Ruderman theorem (on the amplitude, not on the cross section), showing the need of 10 to 20 % correction effects which are indeed obtained in the more general approach. A similar agreement is obtained in the framework of the theoretical models of /84, 86/.

In this context it is interesting to mention that a direct theoretical evaluation of the Breit threshold amplitudes for photoproduction has been performed in the framework of fixed t dispersion relations /88/. These calculations provide for the amplitudes $T_1^{(-)}$, $T_1^{(0)}$, which determine charged pion production, results which are consistent with ours, while there is considerable disagreement for $T_1^{(+)}$, which is not surprising in the light of the previous remarks.

We finally turn to charged electroproduction near threshold which has recently received much attention, both theoretical and experimental, in view of the possibility of an independent determination of $G_\Lambda(t)$, alternative to the direct one by neutrino scattering.

This programme, however, is not so straightforward to achieve. The point is that the measured quantity is the threshold cross section,

$$\sigma = \sigma_T + \varepsilon\sigma_L \propto |E_{0+}|^2 - k^2\varepsilon \left|\frac{L_{0+}}{k_o}\right|^2$$

and, as our previous discussion indicates,

$$E_{0+}^{(-)} \propto \frac{G_A(t)}{2f_\pi} + \text{Nucleon term} + \delta E_{0+},$$
$$\frac{L_{0+}^{(-)}}{k^o} \propto \frac{D(t)}{2f_\pi m_\pi} + \text{Nucleon term} + \delta L_{0+}. \tag{3.71}$$

From the theoretical standpoint these quantities are not on the same footing. Indeed, in E_{0+} the main contribution is given by $G_A(t)/2f_\pi$ and all the approaches, which aim at extrapolating the chiral limit to the physical region, agree in stating that the additional corrections do not exceed 10 to 30 %. On the contrary, the various models disagree in predicting L_{0+}, already in the way the rapidly varying pion pole is introduced. This means that for a more reliable determination of $G_A(t)$ the quantity E_{0+} only should be used, which requires the experimental separation of transversal and longitudinal cross sections at threshold.

Since at present such information is available for a few experimental points only, $G_A(t)$ is practically extracted from the complete threshold cross section and reflects, therefore, a dependence on the theory one is using. We defer to Section 5.4 a more detailed analysis of the experimental results according to the various theories.

F) We now turn to a different consequence of the fundamental commutator (3.51). Historically, the sum rules we shall write in the following were obtained by evaluating the completeness sum (3.52) in a special reference frame, the so-called infinite momentum frame.

In the present context it is more convenient to write first the soft-pion theorems, which follow from (3.62) and (3.64) as $m_\pi^2 \to 0$, $k^2 \to t$. These are

$$T_1^{(-)}(\nu=0,\ k^2=t) = \frac{1}{2f_\pi}\left[G_A(k^2) + \frac{k^2 G_A(0)}{4m_N^2-k^2}\, G_M^V(k^2)\right] \tag{3.72}$$

$$\left(\frac{T_2^{(+,0)}}{\nu}\right)_{\substack{\nu=0\\ t=k^2}} = \frac{G_A(0)}{f_\pi}\ \frac{G_E^{(v,s)}(k^2)}{4m_N^2-k^2} \tag{3.73}$$

One then expresses these low-energy limits through *unsubtracted* dispersion relations at fixed $t = k^2$, $q^2 = 0$ for the relevant amplitudes. To this aim the amplitudes $T_{1,2}$ must be related to the $M_{1\ldots8}$ ones of the general expansion (3.15). Writing for them the dispersive representation and enforcing the soft-pion constraints, one finds, after careful selection of the pole terms and use of the Goldberger-Treiman relation, the following sum rules /92/

$$\frac{G_A(t)}{G_A(0)} = F_1^{(v)}(t) + \frac{4m_N}{g_{\pi N}}\ \frac{t}{\pi}\int_0^\infty \frac{d\nu'}{\nu'}\ \mathrm{Im}\ M_8^{(-)}(\nu',t) \tag{3.74}$$

$$\frac{F_2^{(v,s)}(t)}{2m_N} = \frac{4m_N}{g_{\pi N}}\ \frac{1}{\pi}\int_0^\infty \frac{d\nu'}{\nu'}\ \mathrm{Im}\left[M_5^{(+,0)}(\nu',t) - m_N\, M_6^{(+,0)}(\nu',t)\right]. \tag{3.75}$$

The relevant imaginary parts are of course evaluated at the specific soft-pion configuration $q^2 = 0$, $k^2 = t$, and their asymptotic behaviour guarantees the convergence of the integrals.

These sum rules, a particular and well-known example of a large class of relations derived years ago in the framework of current algebra and PCAC, offer in principle the possibility of a direct evaluation of the interesting nucleon parameters $G_A(t)/G_A(0) - F_1^{(v)}(t)$ and $F_2(t)$ in terms of dispersive integrals. Their evaluation, using experimental inputs for the imaginary parts, is not easy, especially for the sum rule (3.74), and in any case is limited to the small t region. (The required extrapolation from $q^2 = 0$ to $q^2 = m^2$ is not expected to give large errors and, anyway, simple recipes exist for taking it into account).

Careful numerical evaluations have been performed for both sum rules /93/, using improved theoretical and experimental knowledge of the required multipoles at different k^2. We reproduce here the results for the sum rules (3.75) at $t = 0$, i.e., for $F_2^{(v,s)}(0)$, and for the sum rule (3.74) after differentiation at $t = 0$, i.e. for

$$G_A'(0)/G_A(0) - F_1^{(v)\prime}(0).$$

In the first case the relation has been checked with very good accuracy by several calculations which give

$$F_2^{(v)}(0)\Big|_{th.} \approx 3.50 - 3.90,$$

$$F_2^{(s)}(0)\Big|_{th.} \approx 0,$$

to be compared with

$$F_2^{(v)}(0) = 3.70,$$

$$F_2^{(s)}(0) = 0.12.$$

The dispersion calculation of $M_8^{(-)}$ suffers from strong cancellations but it leads, however, to the correct order of magnitude. Using a dipole fit for $G_A(t)$ and the measured value $F_1^{(v)\prime}(0) \approx 0.045\ m_\pi^{-2}$, one obtains for the axial vector "mass",

$$M_A \approx 8\ m_\pi \ \dots\ 9\ m_\pi$$

in agreement with the previous indications.

An alternative possibility for estimating the continuum integrals is represented by the use of an isobaric model for the relevant imaginary parts, which corresponds to the saturation of the sum rules by single particle states of zero width. Although such an evaluation of dispersive integrals may not be completely adequate, it offers, on the other hand, the advantage of reducing the sum rules to a set of *algebraic* relations between hadron parameters, which is useful for a first-hand estimate. In particular, inclusion of the Δ-resonance only gives

$$F_2^{(v)}(0) \approx 4, \; F_2^{(s)}(0) \approx 0,$$

still in the right ball park.

4. Main Features of the Experiments, Preliminary Tests and Measurements

4.1 The Experimental Methods

Electroproduction experiments in the low-energy region are usually divided into two classes: single-arm and double-arm or coincidence experiments. In single-arm experiments only one of the three particles present in the final state is detected. This is usually the lepton, because from the energies l_{01} and l_{02} of the incident and scattered lepton and its angle of scattering θ_1 one derives - by means of (A.4), (A.13), and (A.29) - k^2, W, and ε. No other information can be obtained on the dynamics or on the nature of the other particles. The detection of only one of the recoiling hadrons - instead of the scattered lepton - is never used because it would provide poorer information on the kinematical conditions of each event.

In double-arm experiments one of the two hadrons present in the final state is observed in coincidence with the scattered lepton. Recalling that one event of a reaction of type (1.1) is completely specified by five kinematical variable (Appendix A.2), it is clear that in order to know completely the final state of the hadron system it is sufficient to determine - besides l_{01}, l_{02} and θ_1 - for example, the angles θ_π^* and ϕ_π^* (or θ_N^* and ϕ_N^*) that define the direction of motion of the pion (or of the recoiling nucleon).

Furthermore, the use of coincidences implies the choice of a detector appropriate to the nature of the observed hadron and thus automatically allows a clear separation of the events due to one of the specific reactions (1.1) from the others, or from the "background" (Section 4.1.1).

For values of W not too far from threshold ($W_{th} = m_N + m_\pi$) it is convenient to describe reaction (1.1) as the production by the virtual photon of an intermediate hadron of four-momentum $p_x = k + p_1$ (and mass W) which immediately decays into a pion and a final heavy hadron. The direction of $\underline{p}_x$ [(A.15)] is the axis of the cone inside which the final heavy hadron is emitted with two possible momenta in each direction [(A.18)]. For values of W not too far from threshold, the experimenter can design a heavy hadron detector, which covers the whole emission cone, the semi-aperture of which is given by (A.17).

Above threshold the pion is always emitted in a wide solid angle which very soon extends over 4π. This is the main reason in favour of detecting the neutron recoiling from reaction (1.1a) instead of the pion, in spite of much lower efficiency of the corresponding detector.

4.1.1 The Wide Angle Bremsstrahlung

The background is due to the so-called wide angle bremsstrahlung (WAB) produced in the reaction

$$e + p \rightarrow e' + p' + \gamma, \tag{4.1}$$

the cross section of which is known with good accuracy /94/ and, therefore, can be used as a term of reference for the determination of the electroproduction cross section.

At the energies of interest in the present report the contribution to reaction (4.1) originating from the proton represents always a small correction to the main effect produced by the electron and therefore it can be neglected in the qualitative considerations given below.

To the lowest order (α^3), the photon can be emitted by the electron before as well as after the collision (Fig. 4.1). For fixed values of l_{01}, l_{02}, and θ_1 the photon energy k_0 is fully determined within an appropriate interval, by the energy resolution Δl_{02} of the lepton arm.

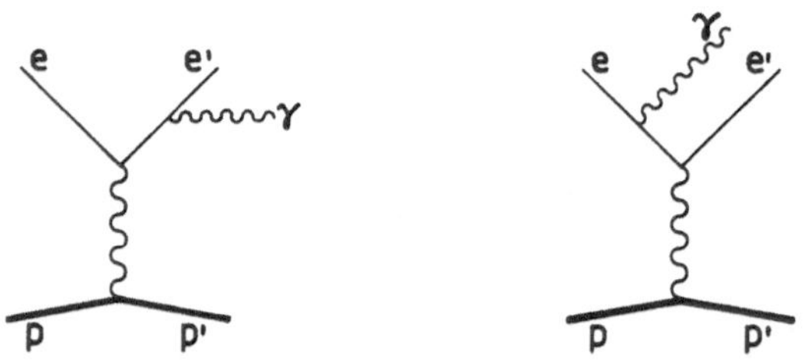

Fig. 4.1. The lowest order (α^3) diagrams which contribute to the wide angle bremsstrahlung

For photons emitted before the elastic collision, one has

$$k_0^b = l_{01} - l'_{01},$$

where

$$l_{01}' = \frac{l_{02}}{1-(l_{02}/m_N)(1-\cos\theta_1)}$$

is the energy that the electron should have before the ep collision in order that after the scattering at the angle θ_1 its energy has the observed value $l_{02} \pm \frac{1}{2}\Delta l_{02}$. For photons emitted after the ep elastic collision the photon energy is given by

$$k_o^a = l_{02}' - l_{02},$$

where

$$l_{02}' = \frac{l_{02}}{1+(l_{02}/m_N)(1-\cos\theta_1)}$$

is the energy that the electron should have after the ep collision in order that with the emission of the photon k_o^a it reaches the observed energy $l_{02} \pm \frac{1}{2}\Delta l_{02}$. The photons k_o^b, k_o^a are emitted in a very narrow cone around the direction of $\underline{l}_1$ ($\underline{l}_2$), the aperture of which, in the so-called peaking approximation, is taken as null.

The cross section is obtained by squaring the sum of the amplitudes of these two processes. The interference term is usually rather small so that for qualitative considerations it can be neglected.

Since the spectra of both types of photons k_o^b, k_o^a are of the bremsstrahlung type, we recognize that the angular distribution of the protons due to reaction (4.1) shows two maxima close to each other and to the angle θ_p^{el} of the protons recoiling from ep elastic collision. By decreasing θ_p starting from $\theta_p > \theta_p^{el}$, the elastic peak is found at θ_p^{el}, the low-angle side of which is increased by the emission of photons of type k_o^a. A second maximum very close to the first is due to the photons of type k_o^b. Not far from electroproduction threshold (for example, $q^* \approx 60$ MeV/c) the total cross section of (4.1) for $l_{01} \approx 1$ GeV is typically 10 times greater than the electroproduction cross section at threshold. The corresponding counting rate observed in a counter of fixed dimensions reaches, however, a rather low value in the angular region where the protons (neutrons) of reactions (1.1b and c) or (1.1a and d) are observed. For values of l_{01} much greater than ≈ 1 GeV, the angular region of the WAB overlaps in part or even completely the angular region of the nucleon recoiling from (1.1).

4.1.2 Typical Double-Arm Experimental Setups

The two main parts of all experimental setups are: a) The electron arm composed of a magnetic spectrometer, Cerenkov and/or shower counters for the identification of the electron, and counter hodoscopes or wire chambers for the reconstruction of the electron trajectories. The spectrometer is usually mounted on a platform that can be rotated around the target centre. The data provided by this setup allow the determination of k^2, W, ε, the single-arm cross section, and the direction $\hat{k}$ of the virtual photon in the l.f.

The main features of the spectrometers and electron detection systems used by three groups working in this field are shown in Table 4.1 which illustrates the status of the art in the seventies.

b) The hadron arm designed to detect one of the hadrons present in the final state.

The determination of this particle's emission angles θ and ϕ is sufficient for all purposes. The relative timing of the coincident hadron and electron is sometimes measured and provides additional very useful information which allows a discrimination between events due to electroproduction and background events.

The detected hadron can be a charged pion, a neutron, or a proton, depending in part on the design of the apparatus, but mainly on the goal of the experiment.

In case of reactions (1.1a) and (1.1c), either the pion or the nucleon can be detected. Usually the pion is detected when the main goal of the experiment is the determination of $F_\pi(k^2)$ (Section 5.2). Then the hadron arm consists of a magnetic spectrometer with a counter and spark chamber arrangement very similar to that used for the electron arm.

The detection of the heavy hadron instead of the pion has the advantage that not too far from threshold this particle recoils only within a cone of semi-aperture given by (A.17), around the direction (A.15). Under these conditions the nucleon detector can be designed to cover the entire emission cone, and the whole angular distribution in θ_N^* and ϕ_N^* can be measured at the same time with considerable advantages from the point of view of both machine time and stability of the detection system. The total electroproduction cross section is obtained by integrating the observed counting rates with respect to θ_N^* and ϕ_N^*.

If only the total cross section is desired, a single nucleon counter can be used, which should be sufficiently large to cover the whole cone of the recoiling particle.

The detection of the nucleon, instead of the pion, has, however, the disadvantage that the coincidence counting rate includes a considerable background originating from the WAB (Section 4.1.1). Such a background can be eliminated by adding to the experimental setup consisting of the electron- and hadron arm a third element indicated as WAB telescope.

This consists of a range telescope designed and placed at such an angle to detect a large fraction of the protons recoiling from reaction (4.1). The output of the WAB telescope is used in active veto in the master trigger of the experiment.

The advantage of the use of the veto from a WAB telescope and of the use of electron-neutron coincidence with respect to single-arm experiments is shown in Fig. 4.2 taken from the work of a DESY group /98/ on reaction (1.1a).

Table 4.1. Main features of the electron arm of typical electroproduction double-arm experiments

	Cornell /95/	NINA /96/	DESY /97/
Energy of the incident electrons (GeV)	9.6	1.5 - 3.5	7.5
Liquid H_2 target (cm)	13	10	10
θ_1		13^0 - 17^0	8.5^0
Momentum acceptance	± 20 % ± 50 %	8 %	5 % - 11 %
Momentum resolution		± 0.3 %	± 0.3 % - ± 0.6 %
Solid angle (sr)		0.6	0.435
Angular acceptance	± 0.5^0 hor. ± 1^0 vert.	1^0 hor. 2^0 vert.	24 mrad hor. 18 mrad vert.
Angular resolution		± 0.1 hor. ± 0.2 vert.	
Total length spectrometer (m)	10	12	40
Identification of electron { gas Cerenkov shower counter	freon lead-lucite	CO_2 lead-perspex	CO_2
Electron trajectory reconstruction	6 spark chambers + scintillation counters	6 scintillation counter hodoscope + 5 scintillation counters	counter hodoscopes + scintillation counters

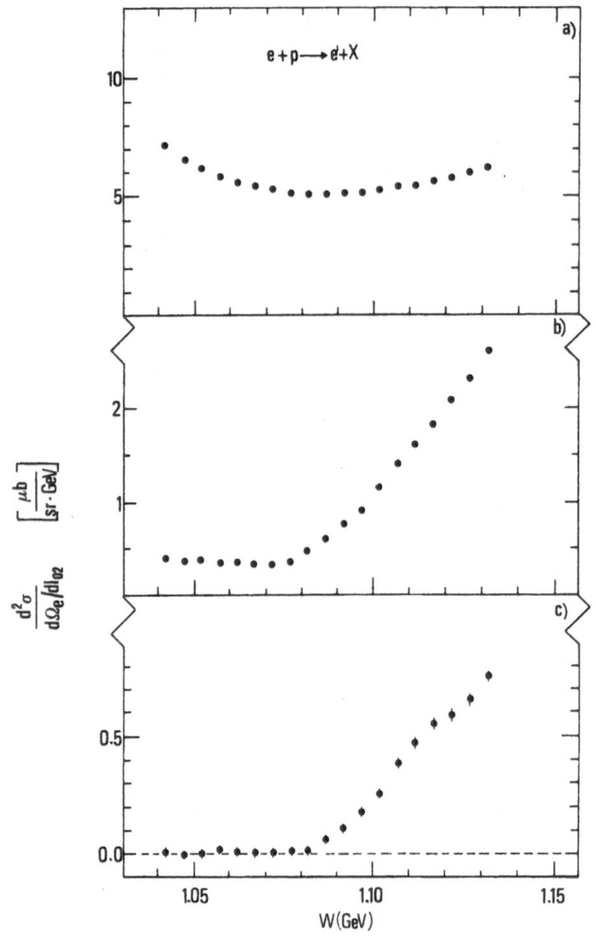

Fig. 4.2. (a) single arm inelastic electron-proton cross section; (b) only those electrons of (a) which have no coincident signal in the WAB telescope are counted; (c) only those electrons of (b) which are in coincidence with a neutron are counted (DESY /98/)

The single-arm inelastic electron-proton cross section is shown in Fig. 4.2a as a function of W. In Fig. 4.2b only those electrons of Fig. 4.2a which have no coincident signal in the WAB telescope are counted. Finally, in Fig. 4.2c only those electrons of Fig. 4.2b which are in coincidence with a neutron are taken into account. The background is completely eliminated so that the threshold of reaction (1.1a) becomes evident. We shall come back to these experiments in Section 5.4.

For moderate values of k^2 and W the angular region where the WAB intensity is maximum falls well outside the cone of the nucleons recoiling from reaction (1.1) only if l_{01} is rather small ($l_{01} \approx 1$ GeV). When l_{01} is increased, keeping k^2 and W constant, the two angular regions tend to overlap so that the WAB telescope should be placed between the liquid H_2 target and the nucleon detector (Sec. 5.4).

4.1.3 The Main Corrections

The momentum calibration of the spectrometers as well as the conversion of counting rates into absolute cross sections are usually obtained by comparing the observed counting rates due to the ep elastic peak with the corresponding theoretical cross section with radiative corrections folded in.

The main corrections that should be applied to the inelastic scattering data include: 1) target wall background; 2) absorption of the detected hadron in the materials crossed along its path; 3) in the case of detected pions, losses for their decay; 4) electron reconstruction efficiency; 5) accidental vetoing of trigger by WAB; 6) trigger efficiency and 7) radiative corrections to inelastic scattering.

The radiative corrections are particularly important: they are different for single-arm and double-arm experiments. For each energy interval between l_{02} and $l_{02} \pm \frac{1}{2}\Delta l_{02}$ the radiative corrections are the difference between two opposite terms: a) an increase due to electrons scattered inelastically with an energy $l_{02}' > l_{02} + \frac{1}{2}\Delta l_{02}$ which enters the interval Δl_{02} because of the irradiation of photons of appropriate energy; b) a decrease due to electrons scattered inelastically with energy in the interval Δl_{02} which are not detected as a consequence of the emission of a photon of sufficiently large energy.

The peaking approximation mentioned in Section 4.1.1 is often adequate for the computation of this correction, which, for example, in the experiments of the Cornell-Harvard group amounts to 30 % /95/ or even 40 % /48/.

The most frequently used papers for their computation are those of MEISTER and YENNIE /99/, and BARTL and URBAN /100/ (by the Cornell-Harvard group), MO and TSAI /101/ (by the Frascati group) and KOHAUPT /102/ and TSAI /103/ (by DESY and NINA groups). The treatise by URBAN /104/ provides an excellent overall presentation of the subject.

4.2 Test of the One-Photon Exchange Approximation

Equations (1.11 - 16) summarize all what can be derived on electroproduction cross section from quantum electrodynamics in the o.p.e.a. The validity of such an approximation can best be tested by checking the ε and ϕ_π^* dependence of the cross section. A less tringent test is provided by the dependence on ε of the measured total cross section, which according to (1.15) should be linear. Almost all the data available today are of this second type. Since they are obtained from single-arm experiments, they refer to the two reactions (1.1a) and (1.1b) together. A few exceptions are mentioned in Sections 5.2.1 and 5.5 in connection with the problem of the separation of τ_L from σ_T for each of the two processes mentioned above, but the intervals of values of the variables ε and ϕ_π^* and/or the accuracy of the measurements are not sufficient for providing significant upper limits for the contributions of two (or more) photon exchange amplitudes.

The linear dependence on ε of the single-arm cross section is a generalization of the Rosenbluth test used in the case of elastic scattering: the Rosenbluth formula is obtained from (1.12) by writing /105/

$$\sigma_T = \frac{4\pi^2}{\alpha k_L} \frac{-k^2}{4m_N^2} G_M^2(k^2)\, \delta\,(1_{02} - 1_{01} + \frac{-k^2}{2m_N^2}),$$

$$\sigma_L = \frac{4\pi^2}{\alpha k_L} G_E^2(k^2)\, \delta\,(1_{02} - 1_{01} + \frac{-k^2}{2m_N^2}),$$

and integrating over 1_{0_2}.

Figure 4.3 shows a plot of the values measured in single-arm experiments of the total virtual-photon cross section,

$$\frac{1}{\Gamma_t} \frac{d\sigma}{d1_{02} d\Omega_1} = \sigma_T + \varepsilon\sigma_T, \tag{4.2}$$

versus ε, for W = 1230 MeV and a few values of $-k^2$. The dots represent the results of LYNCH et al. /106/ working at CEA, the circles /107/ and crosses /108/ the results of two groups working at DESY.

The straight lines represent least square fits to (4.2). The experimental errors are rather large and the points are obtained from different experiments so that the test is not very conclusive, but it is certainly compatible with a linear behaviour in ε.

In the case of elastic scattering, extensive tests of different types have been made, which allow the determination of an upper limit for the two-photon exchange amplitude of about one percent of the one-photon amplitude up to four-momentum transfers of at least 5 $(GeV/c)^2$. Through various indirect arguments, it is reasonable to expect that the one-photon approximation should hold also for electroproduction within the same limits of accuracy established for elastic scattering, at least in the region of very low energy and four-momentum transfer considered here; but more accurate coincidence experiments are highly desirable to test directly this point, on which are based all present theoretical approaches and, consequently, all the analyses of the experimental data. For the time being, the one-photon exchange approximation is accepted as adequate, and the data of the type of those of Fig. 4.3 are used for separating σ_L from σ_T.

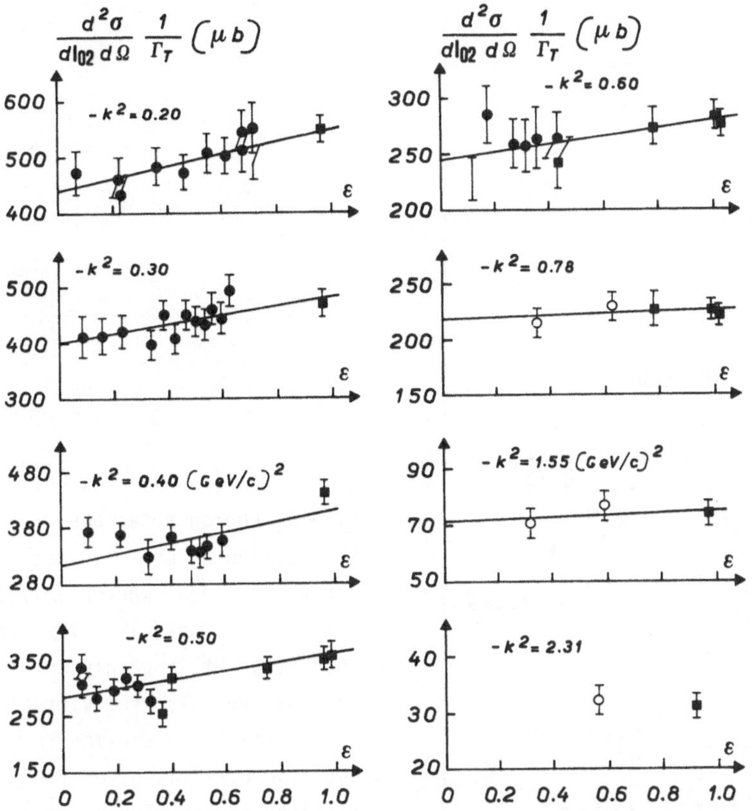

Fig. 4.3. Virtual photon cross section, obtained in the one-arm experiment /108/ as a function of the polarization parameter ε. The linear behaviour is in agreement with o.p.e.a.

4.3 Virtual Photon Total Cross Section

The total electroproduction cross section has recently been measured in the resonance region by various groups, two of which have collected high statistics data with hydrogen and deuterium targets in wide ranges of values of k^2 and W /109, 110/.

In Fig. 4.4 we show the total cross section for absorption of photons by protons

$$\gamma + p \rightarrow \text{anything}$$

for both $k^2 = 0$ (real photons) /111/ and $-k^2 = 1.0$ (GeV/c)^2 /112/ in the energy region of interest in the present report. The figure shows, for comparison, also the virtual photon total cross section for production of neutral pions /113, 114/

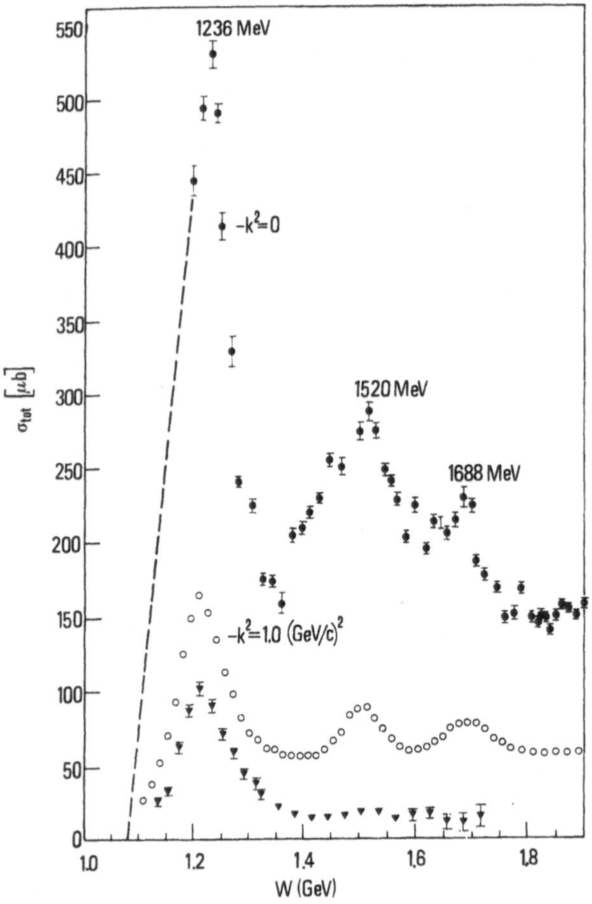

Fig. 4.4. Photon total absorption cross section by protons versus W for $k^2 = 0$ (real photons) /111/ and $-k^2 = 1.0$ /GeV/c$)^2$ /112/. The dashed straight line only to guide the eye. Also the virtual photon total cross section for production of π^0 is given for comparison (triangular points) /113, 114/

$$\gamma_v + p \rightarrow \pi^0 + p$$

deduced by integrating over the angle of emission of the pion the differential cross section obtained from double-arm measurements at $-k^2 = 1.0$ (GeV/c)^2 (Section 5.3).

Notice the decrease undergone by the total cross section passing from real to virtual photons, i.e., after the elimination from the observed cross section of the trivial factor (1.12) appearing in (1.15). Such a trend is particularly marked at the Δ(3.3) resonance (Table 4.2).

Table 4.2. Dependence on $-k^2$ of the total absorption cross section of virtual photons against protons for W = 1232 MeV /111, 112/

$-k^2$ (GeV/c)^2:	0	0.5	1.0	2.0	3.0	4.0
σ_{tot} (μb) :	526	370	160	42	16.	7.5

The values of the resonance mass M_R obtained from seven spectra in the range $-k^2$ = 0.09 $(GeV/c)^2$ to $-k^2$ = 1.82 $(GeV/c)^2$ are all in the interval 1224 to 1236 MeV /110/. The width Γ_R is fairly constant in the covered range of k^2. The highest precision is obtained at $-k^2 = 0.78$ $(GeV/c)^2$: $\Gamma_R = 1.19 \pm 1.6$ MeV /5/.

The ratio of peak to background decreases with increasing momentum transfer /110, 115/.

The deuterium proton total cross section ratio σ_{D_2}/σ_P at resonance is practically constant and very close to 2 up to $-k^2 = 1.5$ $(GeV/c)^2$. This means that there is no indication of an isotensor component of the exchanged photon /109/.

4.4 Separation of σ_L and σ_T from Single-Arm Experiments

Various authors /116, 117/ have separated σ_L and σ_T for both reactions (1.1a) and (1.1b) together from the slopes of plots like those of Figs. 4.3. Figs. 4.5 and 4.6 show the results obtained by BÄTZNER et al. /117/ with a single-arm experiment. The longitudinal cross section remains small compared with σ_T, which shows the well-known resonance behaviour. A tendency of σ_L to be larger at W < 1.2 GeV than at W > 1.2 GeV is observed at all k^2 covered by the experiment. The solid curves are calculated by VON GEHLEN and WESSEL /67/. Varying k^2 at constant W (Fig. 4.6), the longitudinal cross section is significantly larger than zero at W = 1.17 GeV, indicating a flat maximum around $-k^2 = 0.4$ $(GeV/c)^2$, while near W = 1.27, σ_L seems to be rather small at all k^2.

A few other single-arm experiments give the following results for the ratio $R = \sigma_L/\sigma_T$:

$$1.11 \le W \le 1.90 \text{ GeV /118/}, \quad 0.5 \le -k^2 \le 2.0\ (GeV/c)^2, \quad R \le 0.2, \qquad (4.3)$$
$$2.0 \le -k^2 \le 4.0\ (GeV/c)^2, \quad R < 0.35,$$

$$1.495 \le W \le 1.715 \text{ GeV /119/} \quad -k^2 = 1\ (GeV/c)^2, \quad R = 0.20 \pm 0.15, \qquad (4.4)$$

$$\left.\begin{array}{ll} W = 1.9 \text{ GeV /120/} & -k^2 = 0.8\ (GeV/c)^2 \\ W = 2.2 \text{ GeV} & -k^2 = 0.8\ (GeV/c)^2 \end{array}\right\} R - 0.25 \pm 0.15,$$
$$\left.\begin{array}{ll} W = 2.2 \text{ GeV} & -k^2 = 2\ (GeV/c)^2 \\ W = 2.4 \text{ GeV} & -k^2 = 2\ (GeV/c)^2 \end{array}\right\} R = 0.1 \pm 0.1. \qquad (4.5)$$

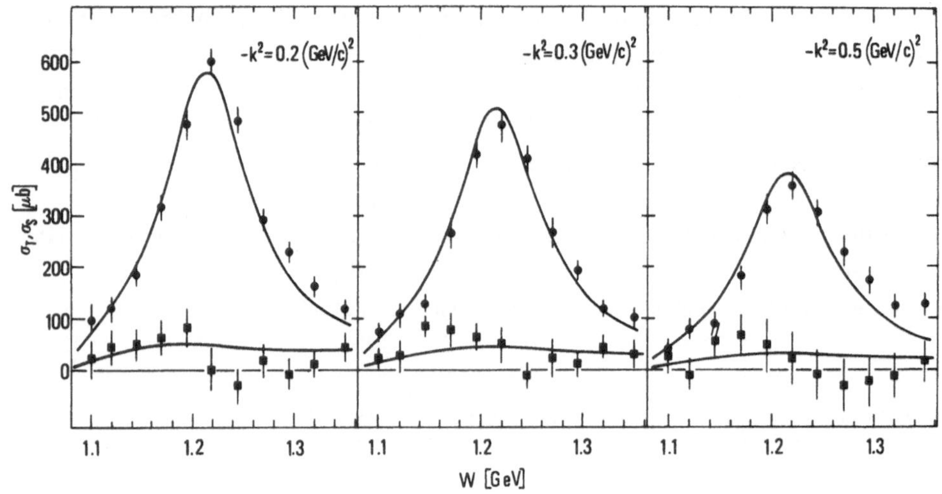

Fig. 4.5. The separated values σ_T (circles) and σ_L (squares) versus W at fixed k^2 according to BÄTZNER et al. /117/. The solid curves are calculations of VON GEHLEN and WESSEL /67/

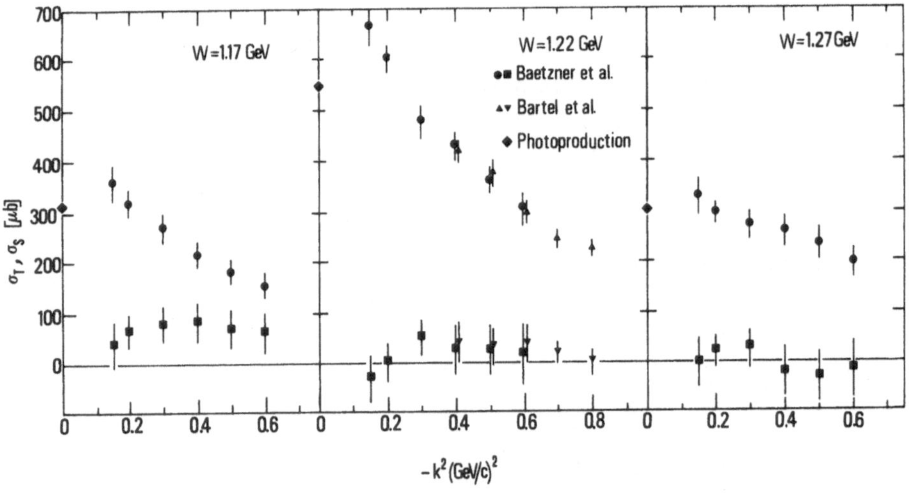

Fig. 4.6. Transversal (σ_T) and longitudinal (σ_L, indicated by the authors as σ_S) cross section as a function of k^2 /116, 117/. The points at $k^2 = 0$ originate from π^+ and π^0 photoproduction measurements

5. Hadron Form Factors from Electroproduction

5.1 The Neutron Electromagnetic Form Factors

The feasibility of deriving the charge form factor of the neutron, $G_E^n(k^2)$, from electroproduction measurements has been considered from time to time as an alternative to the usual method based on the elastic scattering of electrons from the deuteron. While the three other electromagnetic form factors of the nucleons are determined with good accuracy from elastic scattering, the results on $G_E^n(k^2)$ are always affected by rather large errors, as discussed in Section 2.2. For these reasons an independent procedure for determining $G_E^n(k^2)$ would be desirable.

Similar remarks can be applied to the pion form factor, and electroproduction of charged pions has often been suggested as a possible source of additional information on $F_\pi(k^2)$.

The general theoretical framework usually adopted to tackle the problem of extracting such information from electroproduction measurements is represented by fixed-t dispersion relations. Since in the comparison and interpretation of the experimental data one has to separate the polar terms, where the form factors of interest appear as residues of the relevant poles, from the continuum integrals, a - more or less crucial - dependence on the theoretical models is unavoidable. This deals with the evaluation of the dispersive integrals and in particular of the high-energy tails[20] as well as with the choice of the subtraction points, subtraction constants, and so on. On the other hand, looking for suitable kinematical configurations, where the polar terms are expected to represent the leading contribution, can help in reducing the model dependence. In particular a direct determination of $F_\pi(k^2)$ can be tried through an extrapolation of the differential cross section to the pion pole $t = m_\pi^2$. As we shall discuss in the next section, this requires very accurate data.

[20] Which may be partially parametrized by the pion form factor, see the discussion of Appendix D.

As another interesting possibility, we can consider the threshold region where the nucleon pole and, for not too large $|k^2|$, the pion pole very likely dominate the electroproduction amplitude. Then combined measurements of $E_{0^+}(\text{th.})$, $L_{0^+}(\text{th.})$ [i.e., of $\sigma_T(\text{th.})$, $\sigma_L(\text{th.})$], theoretically evaluated in the Born approximation, could in principle provide valuable information on $G_E^n(k^2)$ and $F_\pi(k^2)$. Actually there are some theoretical subtleties since the Born approximation for electroproduction is not automatically gauge invariant[21], unless $F_\pi(k^2) = F_1^V(k^2)$, and the way the pion pole is introduced is not unique. Thus, if one insists on having gauge invariance automatically fulfilled at this level, an ad hoc phenomenological expression must be devised for the longitudinal part, which contains the pion pole.

The usually adopted Born approximation expressions of the threshold multipoles are, for π^+ production /121/

$$E_{0^+}^{\pi^+}(\text{th.})\Big|_{B.A} \propto \frac{1}{2m_N+m_\pi}\left[F_1^p(k^2) - \frac{m_\pi}{2m_N}F_2^p(k^2)\right] - \frac{1}{2m_N^2+m_\pi m_N-k^2}\left[(m_\pi+m_N)\,F_1^n(k^2) + \frac{k^2+m_\pi m_N}{2m_N}F_2^n(k^2)\right], \tag{5.1}$$

$$\frac{L_{0^+}^{\pi^+}}{k_0}(\text{th.})\Big|_{B.A} \propto \frac{1}{2m_N+m_\pi}\left[\frac{m_\pi}{k^2}F_1^p(k^2) - \frac{1}{2m_N}F_2^p(k^2)\right] - \frac{1}{2m_N^2+m_\pi m_N-k^2}\left[\left(1+\frac{m_\pi m_N}{k^2}\right)F_1^n(k^2) + \frac{m_\pi+m_N}{2m_N}F_2^n(k^2)\right] + \frac{k^2-m_\pi^2}{k^2(2m_N m_\pi^2+m_\pi^3-m_N k^2)}F_\pi(k^2). \tag{5.2}$$

Introducing into the expression (5.1) for $E_{0^+}(\text{th.})|_{B.A.}$ the well-known empirical parametrizations of $F_{1,2}^p$ and F_2^n, F_1^n can be obtained from the experimental fit. Substituting the nucleon form factors into $L_{0^+}(\text{th.})|_{B.A}$, $F_\pi(k^2)$, in turn, may be extracted.

Before any conclusion is reached, one must of course estimate the size of the contribution from the continuum integrals. A careful discussion of these background corrections to the Born approximation can be found in the work by VON GEHLEN /67/.

The dispersion integrals were evaluated by inserting the s- and p-wave multipoles (actually their imaginary parts). The most important and best known source of corrections is, in general, the magnetic dipole excitation of the $\Delta(3,3)$ resonance,

---

[21] For the photoproduction process, this difficulty is not present and the dominance of the polar terms is expressed by the approximate validity (within 10 - 20 %) of the Kroll-Ruderman theorem.

although the effect of the second resonance P_{11}[22] can occasionally be important. Also the importance of the high-energy part is hard to estimate.

It turns out that the present uncertainty in the dispersive calculation for $\sigma_T^{\pi^+}$(th.) is larger than the effect of varying $G_E^n(k^2)$ within the limits derived from electron scattering experiments. As a conclusion, measurements of π^+-electroproduction at threshold can hardly provide, at present, a determination of $G_E^n(k^2)$ better than described in Section 2.2.

Background effects as large as 40 % also appear in $\sigma_L^{\pi^+}$(th.), to be used for extracting $F_\pi(k^2)$, so that such a derivation of the pion form factor appears to be affected by large theoretical uncertainties. We devote the next section to a more complete discussion of the $F_\pi(k^2)$ determination.

In connection with the discussion of the corrections to the Born terms, an original approach has been proposed by SUROVTSEV and TKEBUCHAVA /122/. The idea is to find a region of the space of the variables s, t, and k^2 where the cross sections of processes like electroproduction (or photoproduction) are described only by the Born terms.

The differential cross section for a virtual photon can be expressed as the sum of two parts

$$d\sigma/dt = (d\sigma/dt)_{Born} + \Phi(s,t,k^2),$$

where the first is the Born term and the second takes into account the final state interaction and its interference with the Born part of the amplitude. To establish the conditions under which

$$\Phi(s,t,k^2) = 0,$$

the authors use the existence theorem for implicit functions.

According to this theorem if we observe at least one point s_0, t_0, k_0^2, where the effects of rescattering and their interference with the Born terms compensate each other, then, since the cross section is continuous in the physical region, there is a surface of compensation $t = f(s,k^2)$, on which the cross section is given exactly by the Born term only. The intersection of this surface with every plane k^2 = constant defines (compensation) curves in the plane s, t. These curves can easily be constructed in the case of photoproduction $k^2 = 0$ by comparison of the Born cross section with the experimental data. Under the assumption that for electroproduction, at least at small $|k^2|$, the compensation curves are not much different from those

---

[22] $J^P = 1/2^+$, $I = 1/2$, $M \simeq 1434$ MeV, $\Gamma \approx 200$ MeV.

obtained for photoproduction, the authors conclude that along these curves the dependence on the model of the cross section of the process should be reduced to be minimal. In other words, the compensation curves for photoproduction should indicate the optimal experimental conditions for deriving the form factors from electroproduction measurements. This method has been applied for F_π and F_p in the inverse reaction $\pi^- p \to e^+e^- n$ by BEREZHNEV et al. /218/.

5.2 The Charged Pion Electromagnetic Form Factor

A) It was first suggested by FRAZER in 1959 /123/ that the pion form factor could be determined from π^+ electroproduction data. The dependence on $F_\pi(k^2)$ comes in through the one-pion exchange pole diagram and, in order to isolate this amplitude from the others which tend to disguise its effect, FRAZER proposed to extrapolate, à la CHEW-LOW /124/, the pion angular distribution data to the pole at $t = m_\pi^2$, which occurs at an unphysical pion angle. For the determination of the pion angular distribution at fixed values of W and $-k^2$, both the electron and the π^+ must be detected.

The standard procedure is to multiply the pole denominator out of the experimental cross section in such a way that the results lie on a smooth curve, which is extrapolated to the pole to obtain the residue. Thus

$$(t-m_\pi^2)^2 \, d\sigma\Big|_{t=m_\pi^2} = N(m_\pi^2)\, F_\pi(k^2), \tag{5.3}$$

where N(t) is a known function, which for kinematical reasons is proportional to "t". This means that the line of extrapolation is steep in t and small deviations in the data region can lead to large differences at the pole point $t = m_\pi^2$. Very accurate data, beyond the present experimental capabilities, would be required to overcome this difficulty, so that this kind of extrapolation does not provide a reliable determination of $F_\pi(k^2)$. Anyway, if one assumes the parametrization

$$F_\pi(k^2) = \frac{1}{1-k^2/M^2}, \tag{5.4}$$

then M^2 is, by this procedure, bounded to the range /125/

$$0.3 < M^2 < 0.55 \ (\text{GeV})^2.$$

The possibility of applying more sophisticated extrapolation techniques to an analysis of electroproduction data has recently been explored /126/. The conclusions are rather negative, but it is pointed out that the ambiguities inherent to any extrapolation procedure can be reduced by exploiting model-independent constraints of the type of (C.13), even if their validity is for unphysical points. As an application of this idea, the experimental data by PISA-ROME /127/ and DARESBURY-PISA /128/ groups at the electroproduction threshold with $|k^2| < 0.23\ (\text{GeV})^2$ have been extrapolated, together with the $k^2 = 0$ photoproduction value, to the pion pole. Since we move along the threshold line $t = (k^2-m_\pi^2)(1+m_\pi/m_N)^{-1}$ this leads to the time like point $k_\pi^2 = m_\pi^2(2+m_\pi/m_N)$. The value obtained is /129/

$$F_\pi(k_\pi^2) = 1.19 \pm 0.10 \tag{5.5}$$

corresponding roughly to a pion charge radius $\langle r_\pi^2\rangle^{1/2} \approx 0.9...1.1\ \text{fm}^{-1}$.

B) If one has an adequate theory for the complete electroproduction amplitude and if one can find kinematical conditions where the pion pole gives a large contribution, it is possible to fit the theory directly to experimental data to determine the pion form factor. Clearly this approach reduces the burden on the experimenter, but any conclusion about the pion form factor becomes sensitive to the theoretical model used in the fitting.

The models recently employed by various authors are all based on fixed-t dispersion relations but their conclusions about the exact shape of $F_\pi(k^2)$ turn out to be somewhat different, reflecting the different assumptions made in addition to those inherent in dispersion theory. None of the theoretical models employed in these analyses can be considered to be completely satisfactory; nevertheless this approach is at present the only practical way of deriving $F_\pi(k^2)$ from electroproduction. Keeping these liminations in mind, we now proceed to a closer examination of recent and less recent attempts.

The approach was first used by the Cornell University group led by BERKELMAN /14/, who pointed out that $F_\pi(k^2)$ can best be determined at $\theta_\pi^* = 0$, i.e., when the pion is moving in the direction of the incoming virtual photon. The argument is as follows. The $\Delta(3,3)$ resonance that dominates (the imaginary part of) the amplitude is excited mainly by the M_{1+} multipole which contributes to the transverse cross section, not to the longitudinal one. This preliminary remark is clearly confirmed by the experimental results (see Fig. 5.8).

The pion pole amplitude contains the factor $(t-m_\pi^2)^{-1}$ which, in the limit $-k^2 = 0$, is proportional to $(1-\beta^*\cos\theta_\pi^*)^{-1}$. The pole occurs at the unphysical angle $\cos\theta_\pi^* = \beta_\pi^{*-1}$ and therefore the corresponding amplitude should reach its largest value within the physical region at $\theta_\pi^* = 0$.

The transverse contributions, however, never reaches this maximum value because it actually vanishes at $\theta_\pi^* = 0$ owing to angular momentum conservation. A transverse photon has one unit spin angular momentum parallel or antiparallel to its motion (z axis), while the pion moving in the same direction has $j_z = l_z = 0$.

A longitudinal photon, on the contrary, has no helicity to get rid of, so that the longitudinal pion amplitude actually does attain its largest value in the forward direction $\theta_\pi^* = 0$. The Cornell group concludes that the elusive pion pole amplitude should stand out best above the resonance background in the longitudinal part εB of the cross section (1.13) observed at $\theta_\pi^* = 0$. At this pion angle the interference terms in the virtual photoproduction vanish and we have

$$\frac{d\sigma}{d\Omega_\pi^*}(\theta_\pi^* = 0) = A(W,k^2,K) + \varepsilon B(W,k^2,0). \tag{5.6}$$

Furthermore, the longitudinal contribution can be maximized relative to the transverse one by choosing an electron scattering angle θ_1 as small as possible so that the polarization parameter ε is close to 1. Under these conditions one has the further advantage that the electroproduction yield is maximized due to the factor $(1-\varepsilon)^{-1}$ appearing in (1.12). BERKELMAN's qualitative arguments were convincing for the energy region around the $\Delta(3,3)$ resonance ($W \leq 1300$ MeV). A few experiments were made under these conditions /14, 17/. The dispersion theory models, in the meantime, had an increasing success in describing both photoproduction and electroproduction experiments. The confidence in the adequacy of their predictive capacity has increased enough to rely on the conclusion that at $\theta_\pi^* = 0$ the longitudinal pion pole term becomes increasingly dominant as one goes to higher $-k^2$ and higher W. Such a behaviour is clearly shown in Fig. 5.1 taken from /48/. It shows the behaviour versus $-k^2$ of the zero degree ($\theta_\pi^* = 0$) single pion electroproduction cross section for the three values of W used in the experiment of BEBEK et al. /48/, computed with a dispersion theoretic model by BERENDS /130/ adopting two different expressions for the pion form factor.[23] The curve denoted by T gives the contribution of the transverse term (with the dispersion theory corrections included). The rest of the cross section is due to the one-pion exchange diagram. These curves indicate that for modest $-k^2$, the cross section does not decrease as W increases, as a consequence of the increasing dominance of the one-pion exchange. This result suggests that the scattering of the virtual photon by the pion will continue to be a significant part of the total virtual photoproduction cross section even at very high energies.

---

[23] The assumptions of this specific model, by the way, are typical of all such calculations. It is assumed that a) the amplitude is real at high energies where it is represented by the generalized Born approximation and by the dispersion integral, b) the imaginary part is thus given in terms of low-energy data and is dominated by the $\Delta(3,3)$ resonance excited mainly through the M_{1+} multipole. In all these calculations the parametrizations (2.33), (2.34), and (2.35) are used while $G_E^n(k^2)$ is usually taken to be zero.

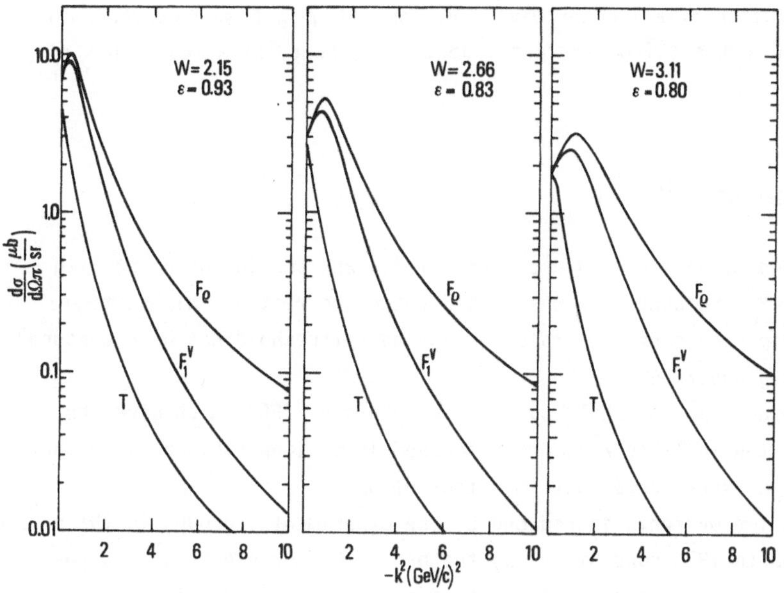

Fig. 5.1. A plot showing the behaviour versus $-k^2$ of the zero degree single pion electroproduction cross section for three c.m. energies W. The curve denoted by F_ρ and F_1^V give the cross section for two popular forms of the pion form factor $F_\pi(k^2)$. The curve denoted by T gives the total contribution to the cross section due to the transverse term including the dispersion theory corrections. The rest of the cross section is due to the one-pion exchange diagram /48/

On the basis of these conclusions some experiments have been made at CEA /131/ and Cornell /48, 132/ at $\theta_\pi^* = 0$, and values of W in the range of 2.15 to 3.11 GeV /133/.

A different approach was proposed and used by the Pisa-Rome group /127/ which suggested taking the measurements near threshold ($W = m_N + m_\pi = 1077$ MeV) where all pions are emitted at $\theta_\pi^* = 0$ and only a few multipoles give appreciable contribution. Near threshold, the production cross section (1.15) reduces to

$$\frac{d^2\sigma}{dT_{02}d\Omega_1} = \Gamma_t \left(\frac{4\pi W|\underline{q}^*|}{m_N k_L}\right) \Sigma + \sigma_B, \qquad (5.7)$$

where

$$\Sigma = |E_{0+}|^2 + \varepsilon(-k^2) \left|\frac{L_{0+}}{k_0}\right|^2 \qquad (5.8)$$

and σ_B is the contribution of all other waves. By computing σ_B from a conveniently chosen model, one can deduce Σ from the cross section measured at a few values of $|\underline{q}^*|/m_\pi << 1$.

$$\Sigma_{th.} = \lim_{\underline{q}^* \to 0} \frac{1}{\Gamma_t} \frac{m_N k_L}{4\pi W|q^*|} \frac{d^2\sigma}{d\Gamma_{02} d\Omega_1} .$$

The results obtained by this second procedure /127, 128/ are in agreement with those obtained at $W \approx M_\Delta$ and above, to which is devoted the rest of this section. The azimuthal angle dependence of the cross section near threshold has been measured by the Daresbury-Pisa group /134/.

The measurements made near threshold by the Saclay group /156/ which give the value of $<r_\pi^2>^{1/2}$ reported in Table 2.1, are discussed in Section 5.4.4 because they are based on a separate experimental determination of σ_L.

C) The most important improvements introduced by the experimental groups in the course of the years from 1968 to 1976 consist of: a) the use of better electron spectrometers and pion arms, b) the extension of the measurements over wider intervals of values of the kinematic variables (in particular θ_π^* and W), c) the collection of better statistics, and d) the use of better theoretical models. These are always along the same general lines and the progress consists substantially in an improved saturation of the dispersion integrals by taking into account resonances higher than $\Delta(3,3)$. Their inclusion has been investigated and shown to have little effect /125/.

For the derivation of $F_\pi(k^2)$, only the data collected in the immediate vicinity of $\theta_\pi^* = 0$ ($\theta_\pi^* \lesssim 3^0$) are always used, in agreement with the original BERKELMAN suggestion. The analysis of the data, in general, is made according to the procedure first used by the Cornell group /14/ which computed the theoretical cross section for each of the experimental points, using the known nucleon form factors [with $G_E^n \cong 0$] and leaving the pion form factor as the only free parameter. For each data point, the value of $F_\pi(k^2)$ which gave agreement with the experiment was determined.

In the following we shall report only on the three most recent experiments /48, 131, 132/ in which two magnetic spectrometers (Table 4.1) were used for detecting the scattered electron and the produced π^+. The analysis of the data is based on a Monte Carlo calculation made assuming unit c.m. cross section and incorporating multiple scattering, detector resolution, and geometrical effects. This procedure enables the authors to correct the number of events in each bin for the acceptance of the apparatus, and hence calculate the observed cross section.

An important novelty introduced in the last work /48/ consists in measuring also the ratio

$$R = \frac{\sigma(\gamma_V D \to \pi^- p p_s)}{\sigma(\gamma_V D \to \pi^+ n n_s)} \tag{5.9}$$

obtained with a D_2 liquid target. The subscript s refers to the spectator nucleon. The knowledge of R allows the separation of the isoscalar virtual photoproduction amplitude a_s from the isovector amplitude a_v, which is the only one containing the pion pole and the Δ contribution.

In terms of isovector and isoscalar virtual photoproduction amplitudes, R is given by

$$R = \frac{|a_s|^2+|a_v|^2-2\mathrm{Re}\,(a_s^* a_v)}{|a_s|^2+|a_v|^2+2\mathrm{Re}\,(a_s^* a_v)} = 1 - 4\,\frac{\mathrm{Re}\,(a_s^* a_v)}{|a_s|^2+|a_v|^2+2\mathrm{Re}\,(a_s^* a_v)}\,, \tag{5.10}$$

so that one can write

$$\begin{aligned} \frac{1}{2}\,(1+R)\,\frac{d\sigma_v}{d\Omega}(\gamma_v p \to \pi^+ n) &= |a_v|^2 + |a_s|^2, \\ \frac{1}{2}\,(1-R)\,\frac{d\sigma_v}{d\Omega}(\gamma_v p \to \pi^+ n) &= 2\mathrm{Re}\,(a_s^* a_v), \end{aligned} \tag{5.11}$$

where $(d\sigma_v/d\Omega)(\gamma_v p \to \pi^+ n)$ is the photoproduction cross section by virtual photon observed in hydrogen. The use of these relations implies that the π^+ electroproduction should be the same with hydrogen and deuterium target. Figure 5.2 shows an example of the comparison of π^+ production observed with hydrogen and deuterium targets. The two cross sections are in excellent agreement and give no indication of a suppression of the deuterium cross section in the forward direction as is observed in photoproduction.

Photoproduction data suggest that in this domain R is a universal function of momentum transfer of the form

$$R = 1 - A\sqrt{|t|}\,, \tag{5.12}$$

where A is a free parameter, that can be obtained by the best fit of the experimental points. Its least square determined value from electroproduction data is

$$A = 0.817 \pm 0.058 \quad (\chi^2 = 38.6;\ \ \mathrm{d.f.} = 49), \tag{5.13}$$

which corresponds to R values somewhat smaller than for photoproduction.

If the relative phases of the isoscalar and isovector amplitudes were known, the authors could estimate a_s^* from the second (5.11) and then subtract $|a_s|^2$ from the previous equation to obtain $|a_v|^2$. Since the relative phase is not known, the authors assume that a_s and a_v are both real or have the same phase. Thus the error in the calculated isovector component of the cross section for the data point with the

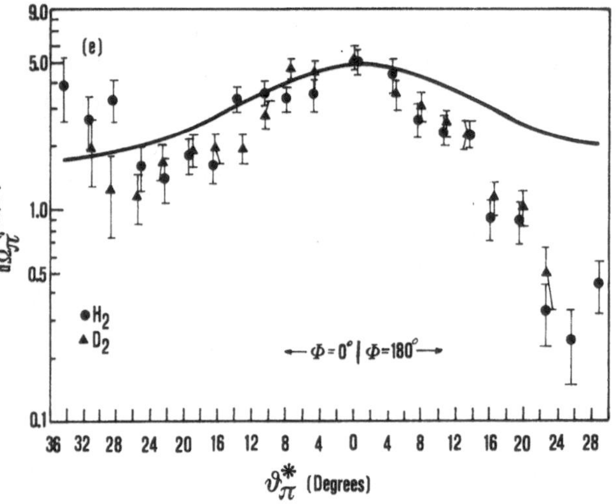

Fig. 5.2. A plot of the virtual photoproduction cross section versus θ^*_π at the nominal values $-k^2 = 1.2$ (GeV/c)^2 W = 2.15 GeV. The solid curve is the prediction of Berends' theory with the empirical addition of the isoscalar component (Cornell-Harvard group: /48/)

smallest value of R (R = 0.436) encountered in the analysis of these experiments is 4 %. This produces roughly a 2 % error in the pion form factor.

Table 5.1 summarizes the isovector components of the cross section and the calculated values of the pion form factor using either R = 1 or the measured value. Table 5.2 is similar to Table 5.1 but refers to previous measurements.

The errors in the pion form factor are only statistical and do not include an estimated overall error in normalization (estimated to be less than 7 %). The correction for the isoscalar component changed the pion form factor by less than 10 % at all points, except the highest $-k^2$ point. At the $-k^2 = 4$ (GeV/c)^2 point the correction for the isoscalar component decreases F_π by 20 %.

Figure 5.3 is obtained from the data of Tables 5.1 and 5.2. It shows a comparison of the values deduces for $F_\pi(k^2)$ from the experimental results of various groups.

Figure 5.3 a shows the uncorrected values, i.e., the values obtained for no isoscalar component. The values deduced recently from a rather accurate single-arm experiment made by a Kharkov group /135/, although slightly higher, are in substantial agreement with those shown in Fig. 5.3 a. Figure 5.3 b shows a plot of values of $F_\pi(k^2)$ obtained with R given by (5.12).

The curves representing $F_1^V(k^2)$, $G_E^p(k^2)$ and

$$F_\rho(k^2) = (1 - k^2/m_\rho^2)^{-1} \tag{5.14}$$

<u>Table 5.1</u>. The values of the pion form factor determined by the Cornell-Harvard group /48/ using the data for $\theta^*_\pi < 3^o$. The uncorrected columns refer to the raw cross section and the pion form factor determined from it. The isovector columns give the calculated isovector component and the pion form factor determined from it. Uncertainties are statistical only.

			Uncorrected			Isovector	
W [GeV]	$-k^2$ $[GeV/c]^2$	$-t$ $[GeV/c]^2$	$\frac{d\sigma}{d\Omega}$ [μb/sr]	F_π	R	$\frac{d\sigma}{d\Omega}$ [μb/sr]	F_π
2.15	1.216	0.069	5.077 ± 0.740	0.324 ± 0.28	0.786	4.535 ± 0.661	0.292 ± 0.026
3.11	1.198	0.019	3.089 ± 0.309	0.321 ± 0.018	0.888	2.916 ± 0.292	0.305 ± 0.017
3.11	1.712	0.034	2.517 ± 0.325	0.257 ± 0.018	0.850	3.328 ± 0.301	0.246 ± 0.017
2.67	3.301	0.162	0.769 ± 0.174	0.136 ± 0.017	0.672	0.643 ± 0.145	0.123 ± 0.015
2.15	1.988	0.157	2.280 ± 0.289	0.221 ± 0.016	0.676	1.911 ± 0.242	0.199 ± 0.015
2.15	3.991	0.477	0.512 ± 0.156	0.124 ± 0.022	0.436	0.368 ± 0.112	0.101 ± 0.019

Table. 5.2. The values of the pion form factor determined from cross sections for $\theta_\pi^* < 3^0$ reported in CEA /131/ and Cornell /132/. The uncorrected columns refer to the raw cross section and the pion form factor determined from it. The isovector columns give the calculated isovector component and the pion form factor calculated from it. Uncertainties are statistical only.

			Uncorrected			Isovector	
W [GeV]	$-k^2$ $[GeV/c]^2$	$-t$ $[GeV/c]^2$	$\frac{d\sigma}{d\Omega}$ [μb/sr]	F_π	R	$\frac{d\sigma}{d\Omega}$ [μb/sr]	F_π
2.15	0.176	0.003	7.15 ± 0.34	0.810 ± 0.044	0.952	6.99 ± 0.33	6.786 ± 0.045
2.15	0.294[a]	0.006	8.05 ± 0.44	0.641 ± 0.028	0.936	7.79 ± 0.43	0.606 ± 0.028
2.15	0.396	0.011	8.90 ± 0.34	0.577 ± 0.016	0.914	8.52 ± 0.33	0.550 ± 0.015
2.15	0.795	0.034	6.99 ± 0.37	0.400 ± 0.013	0.850	6.47 ± 0.34	0.380 ± 0.013
2.15	1.188	0.066	3.54 ± 0.28	0.276 ± 0.014	0.790	3.16 ± 0.25	0.256 ± 0.013
			Cornell '71 /95/				
2.67	0.620	0.011	5.15 ± 0.25	0.465 ± 0.015	0.914	4.93 ± 0.24	0.453 ± 0.014
2.90	1.069	0.019	3.53 ± 0.31	0.323 ± 0.017	0.888	3.33 ± 0.29	0.321 ± 0.017
2.66	1.204	0.031	3.61 ± 0.23	0.291 ± 0.010	0.856	3.35 ± 0.21	0.279 ± 0.010
2.90	1.314	0.048	3.50 ± 0.29	0.266 ± 0.013	0.822	3.19 ± 0.27	0.269 ± 0.012
2.15	1.200	0.069	4.43 ± 0.29	0.288 ± 0.012	0.786	3.95 ± 0.26	0.269 ± 0.011
2.66	2.015	0.070	1.59 ± 0.17	0.185 ± 0.011	0.784	1.42 ± 0.15	0.174 ± 0.010

[a] For this value of k^2, data have been taken also at W = 1.86, 2.05, 2.50 GeV.

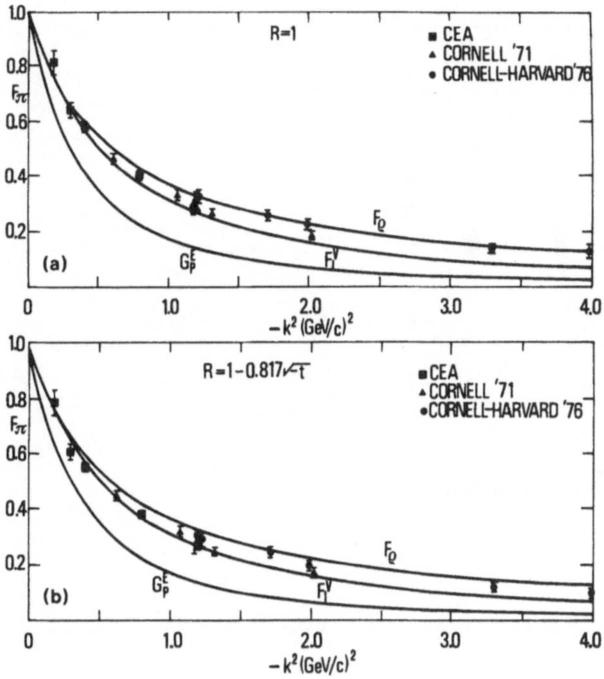

Fig. 5.3. The value of the pion electromagnetic form factor derived from the data using the dispersion theory of Berends: (a) for no isoscalar component and (b) for an isoscalar component given by $R = 1 - 0.817\sqrt{|t|}$. Cornell-Harvard data /48/, CEA data /131/, Cornell '71 data /132/

are also shown for comparison. Figure 5.3 b seems to indicate that $F_\pi(k^2)$ is very close to $F_1^V(k^2)$ up to $-k^2 = 4.0\ (\mathrm{GeV}/c)^2$.

A single-pole expression,

$$F(k^2) = (1 - k^2/m_V^2)^{-1}, \tag{5.15}$$

also gives a good fit to the data with

$$m_V^2 = 0.47 \pm 0.01\ (\mathrm{GeV})^2; \quad (\chi^2 = 20.4,\ \mathrm{d.f.} = 16), \tag{5.16}$$

which is somewhat lower than the square of the ρ-meson mass, $[m_\rho^2 = 0.59\ (\mathrm{GeV})^2]$ (see Fig. 5.4).

The experimental data are now sufficient to see whether the procedure described above gives the same values for $F_\pi(k^2)$ at data points with the same k^2 and different W. The agreement at $-k^2 = 1.2$ and $2\ (\mathrm{GeV}/c)^2$ for three different values of W (2.15,

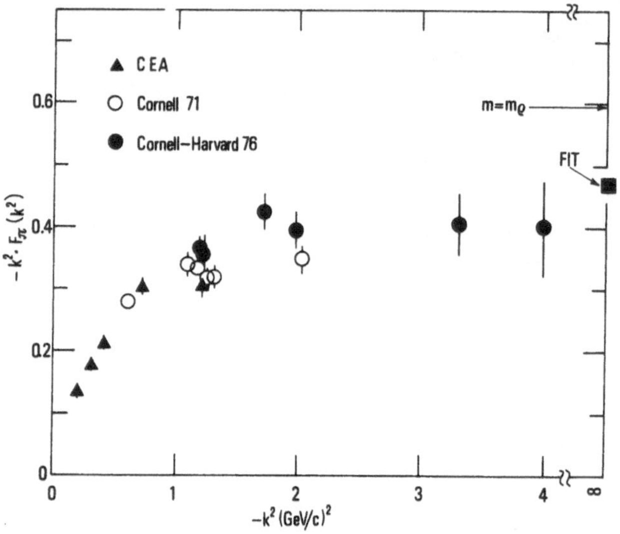

Fig. 5.4. The pion form factor $F_\pi(k^2)$ multiplied by $-k^2$. Data taken from /48, 131, 132/

2.67, 3.11 GeV) is excellent and shows that there is no dependence on the minimum momentum transfer.

Once the pion form factor $F_\pi(k^2)$ and R are obtained from the measurements at $\theta^*_\pi \lesssim 3^0$, the prediction of a given model can be computed over the whole range of variables. The solid curve of Fig. 5.2 is a typical result of such a computation /48/ for the Berends model /130/.

As observed previously /131, 132/, the theory does not correctly predict the longitudinal-transverse interference term D, the transverse-transverse interference term C, or the t dependence at large t.

Another point of interest is the increase in the isoscalar component with $-k^2$ at fixed W, since it has no explanation within the context of the present theory. The isoscalar component observed in photoproduction is explained by the introduction of additional t-channel contributions such as ρ, B, and A_2[24] exchange. In view of the increasing dominance of the pion exchange term at high $-k^2$, this suggests that the exchange of B could give an increasing contribution to the isoscalar component of the cross section. The separation of the longitudinal and transverse cross sections would further ascertain the character of these additional diagrams.

---

[24] A_2: $j^{PC} = 2^{++}$, $I^G = 1^-$, $M_{A_2} = 1310$ MeV, $\Gamma_{A_2} = 102$ MeV
B : $j^{PC} = 1^{+-}$, $I^G = 1^+$, $M_B = 1235$ MeV, $\Gamma_B = 125$ MeV

5.3 Neutral Pion Electroproduction

The experimental investigation of reaction (1.1b) is of considerable interest not only per se, but also because this process should be described by the same models used for reaction (1.1a) with the "one-pion exchange" term left out. This means that the experimental results on reaction (1.1b) provide a unique test of the validity of some of the theoretical assumptions on which the theoretical models are based.

In particular it follows from the absence of the one-pion exchange graph, that the longitudinal cross section σ_L should be small. Therefore one can hope to get some indirect information on the k^2 dependence of the transverse cross section from the comparison of photoproduction and electroproduction of neutral pions.

5.3.1 The Electroproduction of π^0 Near Threshold

The experimental study of reaction (1.1b) has been made near threshold by three groups working at Frascati /127/, DESY /136/ and NINA /137/ using the same experimental setups exploited for the determination of $G_A(k^2)$ from π^+ electroproduction (Section 5.4). The only difference is that, in the present case, instead of the recoiling neutron, the recoiling proton is detected in coincidence with the inelastically scattered electron.

The NINA and DESY groups measured the differential cross section as a function of θ^*_π and ϕ^*_π, which differ by 180^0 from θ^*_p, ϕ^*_p, so that the coefficients $\bar{A}_i$ defined by (1.18) could be determined.

Near threshold, where $W \rightarrow m_N+m_\pi$ and $q^* \equiv |\underline{q}^*| \rightarrow 0$, it is more convenient, however, to express the structure functions appearing in the cross section (1.13) in terms of q^* instead of W [see (A.10)] and develop them in powers of q^*. Assuming that only s- and p-wave multipoles give appreciable contributions, the electroproduction cross section (1.11 - 13) can be recast into the form

$$\frac{d^5\sigma}{dl_{02}d\Omega_1 d\Omega^*_\pi} = \Gamma_t \frac{q^*}{4\pi k_L} (a_0+a_1 q^*\cos\theta^*_\pi + a_2 q^{*^2} \cos^2\theta^*_\pi + a_3 q^* \sin\theta^*_\pi \cos\phi^*_\pi + \tag{5.17}$$

$$+ a_4 q^{*^2} \sin^2\theta^*_\pi \cos 2\phi^*_\pi + a_5 q^{*^2} + a_6 q^{*^2} \sin\theta^*_\pi \cos\theta^*_\pi \cos\phi^*_\pi).$$

The coefficients a_i may be expressed in terms of multipoles using the expressions (C.7). In the o.p.e.a. they are related to the coefficients $\bar{A}_i$ defined by (1.18) by the relation

$$\frac{q^*}{4\pi k_L} a_i = \bar{A}_i \quad (i = 0 \ldots 6)$$

and to the A_i used in /137/ as follows:

$$a_i = A_{i+1} \quad (i = 0 \ldots 6).$$

By integrating the cross section (5.17) with respect to $\cos\theta_\pi^*$, we obtain

$$\frac{d\sigma}{dl_{02}d\Omega_1 d\phi_\pi^*} = \Gamma_t \frac{q^*}{2\pi k_L} (a_0 + bq^{*2} + cq^{*4} + \frac{1}{4}\pi a_3 q^* \cos\phi_\pi^* + \frac{2}{3} a_4 q^* \cos 2\phi_\pi^*), \quad (5.18)$$

where

$$b = \frac{1}{3} a_2 + a_5,$$

and the term cq^{*4} has been added to account for the d-wave contribution. By further integrating with respect to ϕ_π^* we obtain

$$\frac{d\sigma}{dl_{02}d\Omega_1} = \Gamma_t \frac{q^*}{k_L} (a_0 + bq^{*2} + cq^{*4}). \quad (5.19)$$

It is also convenient to work in terms of the slope of the cross section, which can be defined as

$$\frac{4\pi}{\Gamma_t} \lim_{q^* \to 0} \frac{k_L}{q^*} \frac{d^5\sigma}{dl_{02}d\Omega_1 d\Omega_\pi^*} = \lim_{q^* \to 0} (a_0 + bq^{*2}) = a_0. \quad (5.20)$$

Similar definitions applied to (5.18) and (5.19) instead of (5.17) lead to the same value of the slope at threshold. According to (C.13), this is expressed in terms of only two multipoles

$$a_0 = \lim_{q^* \to 0} \left(|E_{0+}|^2 - \varepsilon k^2 \left|\frac{L_{0+}}{k_0}\right|^2 \right). \quad (5.21)$$

The linear behaviour in q^{*2} expected near threshold for the slope (5.20) has been measured by the DESY group /136/ for both the $p\pi^0$ and $n\pi^+$ channel. They found that the ratio of the $p\pi^0$ to the $n\pi^+$ data increases with increasing $|k^2|$.

Figure 5.5 shows the experimental results obtained by the three groups mentioned above, for a_0 and b as functions of $-k^2$ and their comparison with theoretical predictions made by DOMBEY and READ /86/, DEVENISH and LYTH /138/ and BENFATTO et al. /84/ from the weak PCAC model (see Section 5.4). The latter model gives predictions only for a_0, the experimental values of which are in agreement, within two standard deviations, with all three above-mentioned models. They agree also with the results obtained from the Born terms only with pseudovector pion-nucleon coupling (but not with pseudoscalar coupling) /136/.

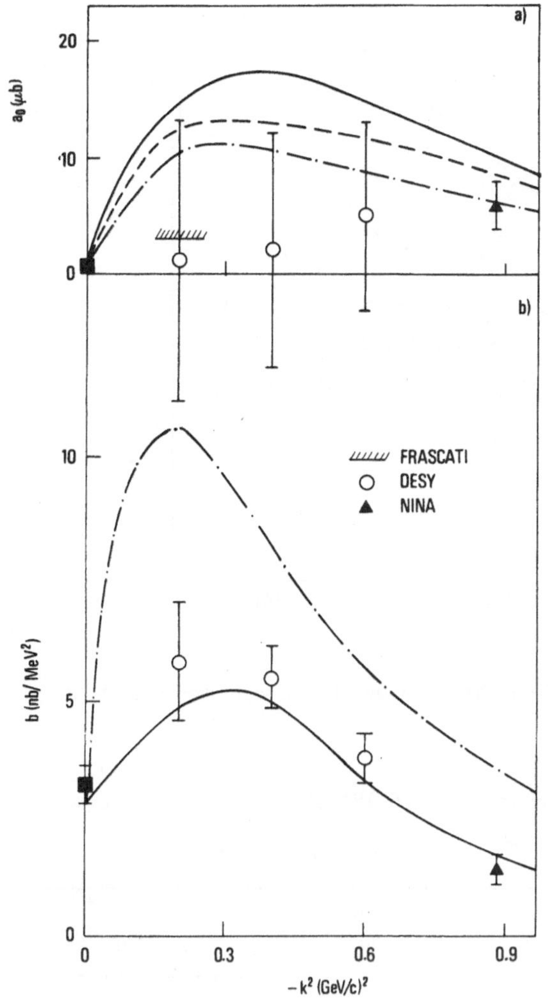

Fig. 5.5. The coefficients a_0,b of the cross section (5.18) as functions of $-k^2$. Experimental points from /127, 136, 137/ —— DEVENISH and LYTH /138/, --- BENFATTO et al. /84/, -·-·- DOMBEY-READ /86/, ■ /139/

The Daresbury-Pisa group /137/ has measured also the other coefficients appearing in (5.18).

Their results (Table 5.3) are in reasonable agreement with the theoretical prediction of DOMBEY and READ /86/ and DEVENISH and LYTH /138/. Figure 5.6 shows the ϕ_π^* integrated data plotted as a function of W with the errors evaluated using the correlations between the coefficients determined by the fit of the Daresbury-Pisa data. Although slightly high, the DEVENISH and LYTH model is much closer to the experimental values over the whole W interval than the DOMBEY-READ model. The lowest-W "data point" from /139/ interpolated between $-k^2$ = 0.6 and 1.0 (GeV/c)^2, also given in the figure, is in fairly good agreement with the results of the Daresbury-Pisa group.

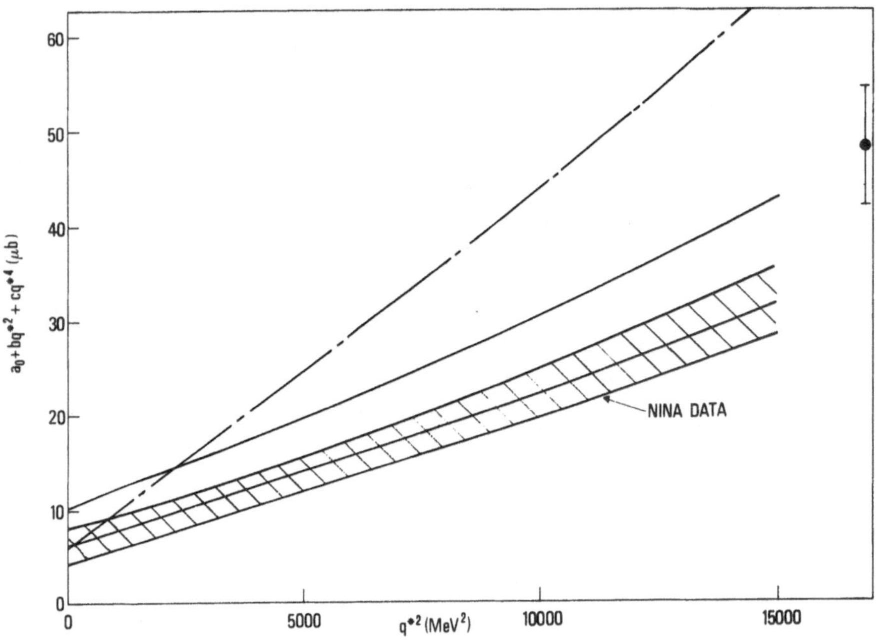

Fig. 5.6. Fit to the total $\pi^0 p$ cross section obtained by the Daresbury-Pisa group /137/, —— /138/, -·-·- /86/, • /139/

Table 5.3. Coefficients appearing in the expression (5.18) of the ep → e'pπ^0 cross section near threshold

	Experimental values NINA data /137/	Theoretical DR /86/	DL /138/
a_0 μb	6.2 ± 2.0		
b(nb/MeV^2)	1.4 ± 0.3		
c(nb/MeV^4)	(1.9 ± 1.3) x 10^{-5}	2.4 x 10^{-5}	1.3 x 10^{-5}
a_3(nb/MeV)	(-0.10 ± 0.08) x 10^{-5}	- 0.22	- 0.25
a_4(nb/MeV^2)	0.13 ± 0.11	+ 0.3	+ 0.4

5.3.2 The Electroproduction of π^0 in the First Resonance Region

A number of coincidence experiments have been made on the electroproduction of π^0 in the region of the first pion-nucleon resonance Δ(3,3) /140/. We shall report mainly on the more recent papers because of the extension and accuracy of their

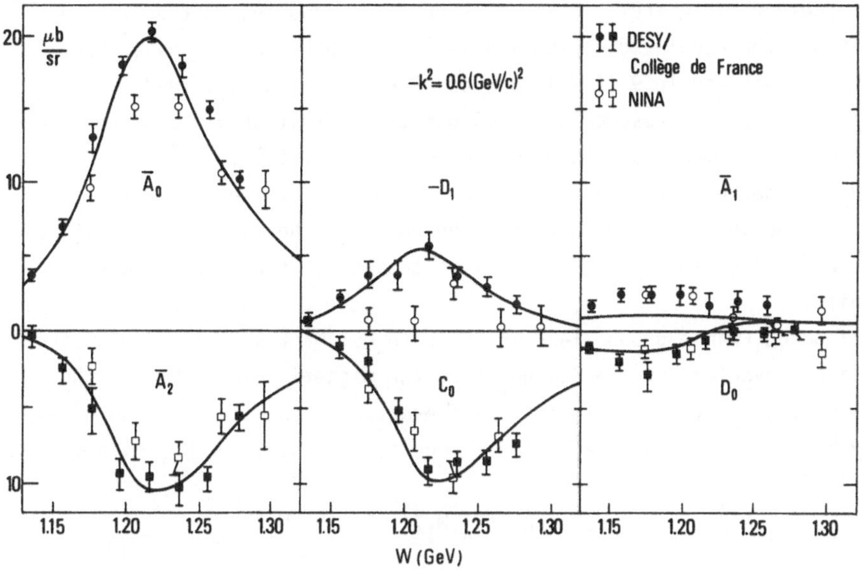

Fig. 5.7. Angular coefficients at $-k^2 = 0.6\ (GeV/c)^2$ compared with dispersion theory of VON GEHLEN and WESSEL /67/

data. Two College de France-DESY groups have measured at $\varepsilon = 0.90$ and $-k^2 = 0.6$, 1.0 $(GeV/c)^2$ /141/ and $-k^2 = 1.56\ (GeV/c)^2$ /142/ the angular coefficients appearing in (1.17, 18) (with d- and f-wave contributions included for $W > 1.565$ GeV.

The angular coefficients, determined by fits of (1.17, 18) to the measured cross section (including contribution of s- and p-waves only) are shown, for $-k^2 = 0.6$ $(GeV/c)^2$ in Fig. 5.7, where the solid curves are computed with dispersion theory /67/. NINA data are taken from /143/.

Pure magnetic dipole dominance would require

$$\overline{A}_0 : \overline{A}_2 = -5 : 3, \quad \overline{A}_0 = C_0, \quad \overline{A}_1 = D_1 = D_0 = 0. \tag{5.22}$$

Clearly the experimental results cannot be described by a magnetic dipole M_{1+} alone.

5.3.3 Interpretation of the Angular Distributions

Although the angular coefficients represent all the information contained in the experimental data, their decomposition in multipoles is useful to provide some idea on the physical content of the measured angular distribution. Furthermore, a few models allow the computation of some definite amplitudes but do not give predictions on cross sections /144/.

The problem, however, is mathematically undetermined. Assuming that only s- and p-waves are present, the angular distribution is described by only 6 measurable coefficients, which increase to 9 if ε is varied. The multipoles involved, however, are 7 with unknown phases, corresponding to 13 parameters that should be determined. In photoproduction, where many more experimental data are available, π^+ and π^0 results are analyzed together, allowing for the decomposition of the multipoles in their 1/2 and 3/2 isospin parts. Furthermore, the phase of the multipole amplitudes of definite isospin are given by the Fermi-Watson theorem in terms of the corresponding πN phase shifts.

The DESY-College de France group /142/ in order to estimate some multipoles from π^0 data only, takes advantage of the dominance of the magnetic dipole M_{1+} in the vicinity of the $\Delta(3,3)$. They assume that the differential cross sections can be explained roughly by $|M_{1+}|^2$ and the five interference terms containing M_{1+}, i.e.,

$$\mathrm{Re}\,(E_{1+}M_{1+}^*),\ \mathrm{Re}\,(S_{1+}M_{1+}^*),\ \mathrm{Re}\,(S_0M_{1+}^*),\ \mathrm{Re}\,(E_{0+}M_{1+}^*),\ \mathrm{Re}\,(M_{1-}M_{1+}^*).$$

Thus the problem is reduced to the determination of a number of unknowns equal to the number of angular coefficients obtained by best fits of the experimental data.

All other terms, which do not contain M_{1+}, are either neglected or computed in some conveniently chosen approximation. For example $|E_{0+}|^2$ is certainly larger than its projection on M_{1+}. Then these authors make use of the recipe

$$|E_{0+}|^2 \approx \frac{[\mathrm{Re}(E_{0+}M_{1+}^*)]^2}{|M_{1+}|^2}\ , \tag{5.23}$$

the importance of which increases off resonance. Similar approximations are used to estimate the other terms.

Figure 5.8 shows the results of such an analysis, which refers to spacelike photons of $-k^2 = 0.6\ (\mathrm{GeV}/c)^2$. [A similar behaviour is exhibited at $-k^2 = 1\ (\mathrm{GeV}/c)^2$]. We see that: a) $|M_{1+}|^2$ shows the expected resonance shape (as in photoproduction);[25] b) $\mathrm{Re}(M_{1-}M_{1+}^*)$ crosses zero near resonance (as in photoproduction); c) $[\mathrm{Re}(E_{1+}M_{1+}^*)/|M_{1+}|^2]$ is negative and of the order of a few percent (as in photoproduction); c) $\mathrm{Re}(E_{0+}M_{1+}^*)$ stays positive throughout the resonance region (while in photoproduction it crosses zero near resonance); e) there is a considerable longitudinal-transverse interference term indicating that $|L_{1+}/M_{1+}|$ is of the order of 5 to 10 %.

The main behaviour of the data is well described by the solution derived by VON GEHLEN and WESSEL /67/. Similar calculations have been performed by CRAWFORD /147/

[25] For π^0 photoproduction data and their analysis see /145, 146/.

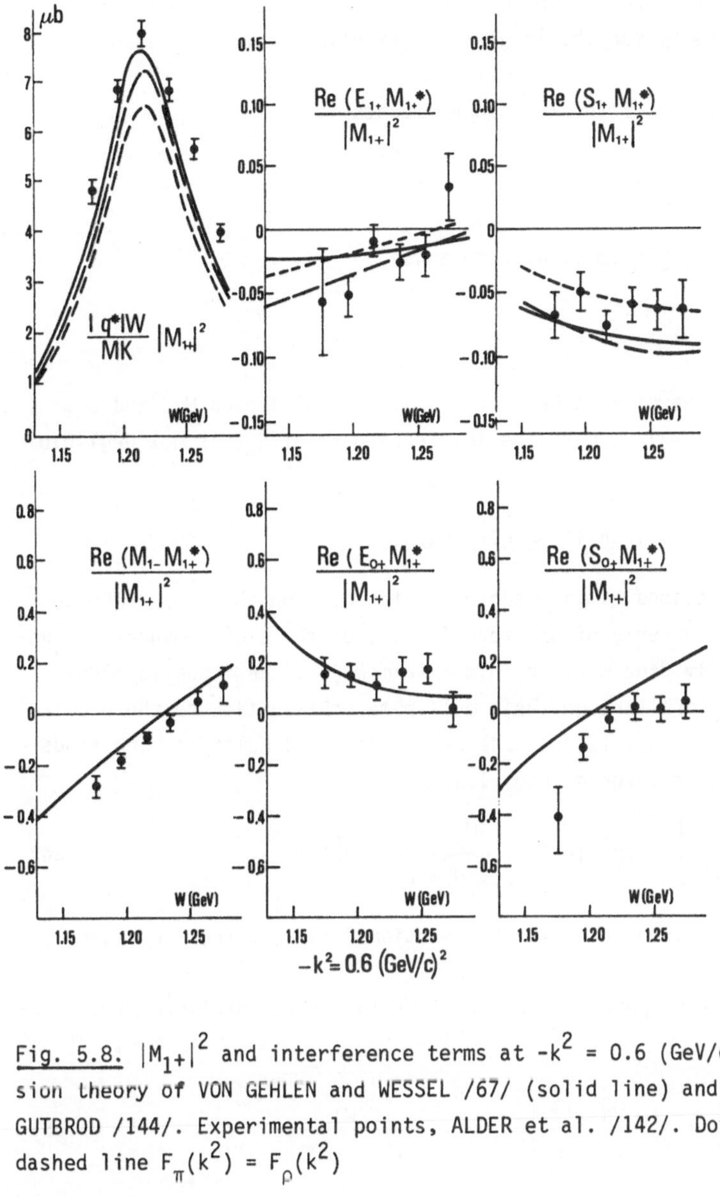

Fig. 5.8. $|M_{1+}|^2$ and interference terms at $-k^2 = 0.6\ (GeV/c)^2$ together with dispersion theory of VON GEHLEN and WESSEL /67/ (solid line) and Bethe-Salpeter model of GUTBROD /144/. Experimental points, ALDER et al. /142/. Dotted line $F_\pi(k^2) = G_E(k^2)$, dashed line $F_\pi(k^2) = F_\rho(k^2)$

while a comparison with the dispersion relation calculations of DEVENISH and LYTH can be found in /144, 148/.

One should also mention the remarkable agreement with a model due to GUTBROD /144/, where the Bethe-Salpeter equation (in ladder approximation) is used to compute the inhomogeneous part of the partial wave dispersion relations for the resonant multipoles $M_{1+}^{3/2}$, $E_{1+}^{3/2}$, and $L_{1+}^{3/2}$.

In conclusion, the main features of the data can be summarized as follows:

1) The magnetic dipole gives by far the largest contribution in the cross section up to $-k^2 = 1.6\ (\mathrm{GeV}/c)^2$.

2) The longitudinal excitation of the resonance amounts to

$$|L_{1+}/M_{1+}| \sim 5 \ldots 10\ \%.$$

3) The resonant quadrupole E_{1+} is small at resonance

$$|E_{1+}/M_{1+}| < 5\ \%.$$

4) Interference of s- and p-waves is clearly visible. The multipoles M_{1-} and L_{0+} contribute considerably at $W < 1200$ MeV. The imaginary part of E_{0+} is not negligible at resonance.

5.3.4 Determination of the N-Δ Transition Form Factor

As discussed in previous sections, it is meaningful to describe the k^2-dependence of the physical multipoles in terms of the form factors of the electromagnetic transition $\Delta \to N\gamma$. This is easily done working in the framework of an isobaric model and expressing the form factors through their multipole contribution to the electroproduction partial wave expansion. In the case of the magnetic multipole one finds after evaluation of the isobar diagram (Fig. 1.3d)

$$M_{1+} = \frac{m_N+M_\Delta}{2m_N^2}\left[\frac{(M_\Delta+m_N)^2-m_\pi^2}{(M_\Delta+m_N)^2-k^2}\right]^{1/2} |\underline{p}^*||\underline{p}'^*| \cdot \frac{M_\Delta g^*}{s-M_\Delta^2+i\Gamma M_\Delta}\ G_M^{(1)}(k^2). \qquad (5.24)$$

The quantities g^*, $G_M^{(1)}$ have been defined in Section 2.1.4 and the kinematical variables refer to the c.m.

A slightly more general (and perhaps more familiar to experimentalists /149/) relation is

$$[G_M^*(k^2)]^2 \equiv \frac{[G_M^{(1)}(k^2)]^2}{[1-k^2/(M_\Delta+m_N)^2]} = \frac{1}{6\pi}\ \frac{m_N^2}{W^2}\ \frac{|\underline{q}^*|}{|\underline{k}^*|^2}\ |M_{1+}|^2\ \frac{\Gamma(W)}{\sin^2\delta_{33}(W)}, \qquad (5.25)$$

where $\delta_{33}(W)$ is the physical $j = I = 3/2$ π-N phase-shift and Γ is the resonance width (allowed to be W-dependent). For $W \to M_\Delta$, $\delta_{33}(W) \to \pi/2$, $\Gamma(W) \to \Gamma_R \approx 120$ MeV, and one can easily reobtain the previous result. Similar relations can be derived for the form factors $G_E^{(1)}$, $G_C^{(1)}$ in terms of the E_{1+}, L_{1+} multipoles.

In Fig. 5.9 one finds the plot of the experimental results for the quantity

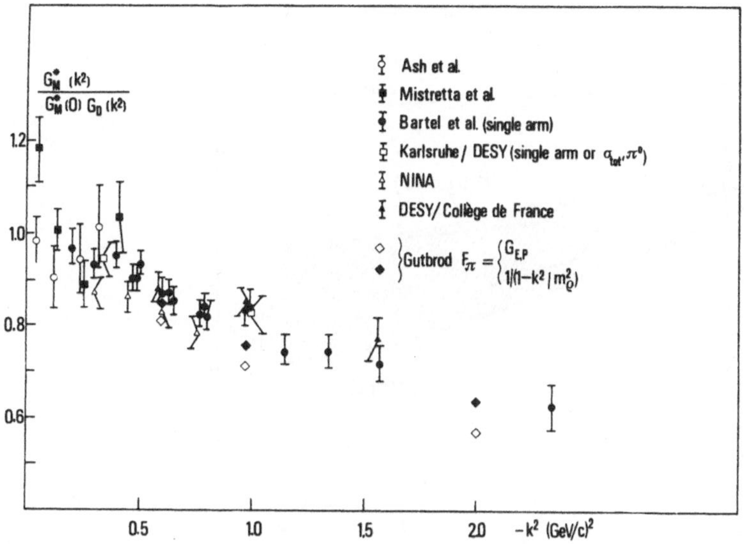

Fig. 5.9. Transition form factor $G_M^*(k^2)$ normalized to $G_M^*(0) = 3$ and to the dipole form factor $G_D(k^2) = (1-k^2/0.71)^{-2}$ /21/

$$R^*(k^2) = \frac{G_M^*(k^2)}{G_M^*(0)} \frac{1}{G_D(k^2)} , \tag{5.26}$$

where $G_D(k^2) = (1-k^2/0.71)^{-2}$ is the typical dipole fit for $G_E^p(k^2)$.

We see that the fall-off for $G_M^*(k^2)$ is more rapid than for $G_D(k^2)$. The form factor $G_M^{(1)} \equiv G_M^*(k^2)[1-k^2/(M_\Delta+m_N)^2]^{1/2}$, does not fall off as rapidly as $G_M^*(k^2)$ but still faster than a dipole. This indication that the "radius" of the $\Delta(3,3)$ is larger than that of the nucleon could be intuitively understood if one considers the Δ as an excited level with a looser structure than the nucleon ground state. In this context it is interesting to mention the possibility of a theoretically motivated fit to the k^2 dependence of the magnetic form factor $G_M^{(1)}(k^2)$ /21/. This has been done in the framework of a general picture for all the electromagnetic transition form factors $N^* \to N\gamma$, with the aim of exploiting a possible correlation between the dynamical k^2 dependence and the spin of the electroexcited resonance. This can clearly have important implications in deriving, for instance, simple rules for the form factor asymptotic behaviour and in throwing some light on different constituent models.

A general analysis of the recent electroproduction data has been performed, in this spirit, in and above the first resonance region ($W \gtrsim 1.4$ GeV) and up to $W \approx 2$ GeV and for $|k^2| < 1(\text{GeV/c})^2$. The theoretical input is a simple parametrization of the various resonance form factors and a Breit-Wigner shape to evaluate the

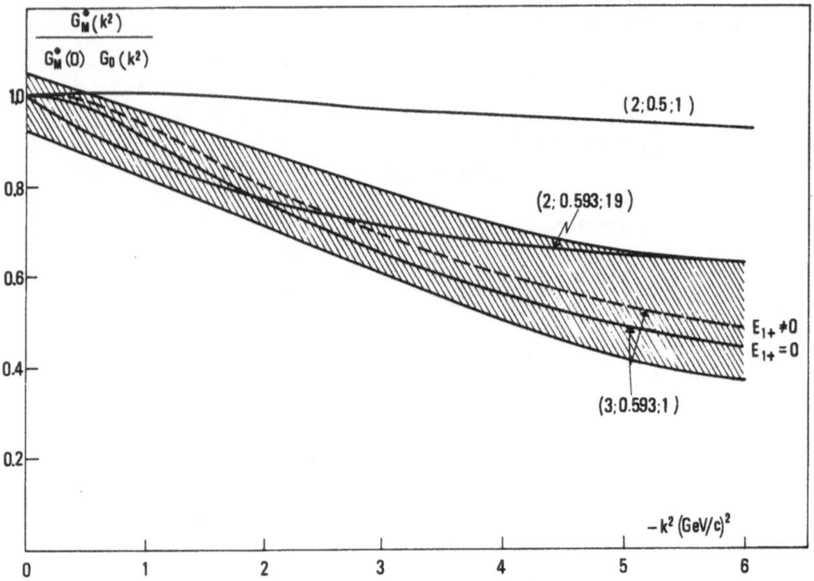

Fig. 5.10. Normalized transition form factor: experimental band compared with theoretical results of /21/ for a few sets of values $(c,\ m_\rho^2,\ \alpha)$

imaginary parts of the amplitudes, while real parts are obtained via dispersion relations. A fit of the data then allows a determination of the input form factor parameters.

In this framework the following simplified formula has been proposed, expressing the ratio $R(k^2)$ as a product of poles:

$$R(k^2) \equiv \frac{G_M^{(1)}(k^2)}{G_M^{(1)}(0)} \frac{1}{G_D(k^2)} = \left[1 - \frac{k^2}{(M_\Delta+m_N)^2}\right] \cdot (1-k^2/0.71)^2 \prod_{n=0}^{c} \left(1- \frac{k^2}{m_\rho^2+nk_0^2}\right)^{-1} \tag{5.27}$$

m_ρ^2 is nearly the ρ-meson mass while k_0^2 and the integer c are free parameters. The situation is depicted in Fig. 5.10, where the dependence on the set of parameters $(m_\rho^2,\ c,\ k_0^2)$ is illustrated. The best fit is obtained with $m_\rho^2 \simeq 0.593$ and $c = 3$, $k_0^2 = 1$.

5.4 The Weak Form Factors of the Nucleon

Threshold electroproduction of positive pions has received considerable experimental attention in view of the possibility of determining the axial vector form factor of the nucleon. Some of the theoretical ideas behind this expectation have been examined in Section 3. We shall discuss later the interpretation from this point of view of the experimental results and the limits of validity of the approach, devoting the first part of this section to a description of the experimental aspects.

5.4.1 The Experiments

Three groups have made coincidence experiments on π^+ electroproduction, at Frascati, /127, 150/, DESY /98/, and NINA /128, 151, 152/, in the frame of the considerations presented above. In all cases, besides electron and neutron arms, the experimental arrangement includes a WAB telescope (Section 4.1.1).

In the experiment of the group working at Frascati, the neutron detector consists of a single liquid scintillation counter, designed to cover the whole solid angle allowed by the kinematics of the reaction, as long as W is close to threshold.

Under these conditions, the counting rate is proportional to the electroproduction cross section integrated over the solid angle of the emitted π^+.

A scintillation counter and an absorber placed in front of the neutron counter allowed the elimination of the proton due to reaction (1.1b) and of those belonging to the tail of the angular distribution of reaction (4.1). The absorber also strongly reduced the low-energy background. The counting rate of the WAB telescope was used to normalize the electroproduction data and thus to derive the absolute value of the cross section.

The magnetic spectrometer selected five (later six) channels corresponding to certain energy intervals Δl_{0i} (i = 1 to 6) very close to each other. The spectrometer was adjusted so that one (or two) of these channels were below, while the others were above, the electroproduction threshold. The channels below threshold detected only the electrons due to background processes, while the others were fired also by the electroproduction events. The cross section at threshold was determined at $-k^2$ = 0.16 (0.84); 0.20 (0.74); 0.24 (0.81) $(\mathrm{GeV}/c)^2$ (ε).

The other two groups working at DESY /98/ and NINA/151/ have both used a high-resolution spectrometer in the electron arm with hodoscopes for the reconstruction of the electron trajectory (Table 4.1). For the neutron detector they used arrays of plastic scintillation counters which allow the determination of θ_n and ϕ_n in the laboratory frame and therefore, by the appropriate Lorentz transformation, the angular distribution of the emitted pions in the c.m. of the π^+n system. Thus both groups separated the four terms appearing in (1.13).

Because of the rather large energy l_{01} of the incident electrons, the WAB telescope had to be placed within the cone of the electroproduced neutrons. Therefore an important correction had to be applied to the counting rates originating from the inefficiency of the WAB telescope.

In the DESY experiment the scattered electrons were grouped by the counter hodoscope according to the missing mass W. The bin in W was 5 to 9 MeV and the overall acceptance changed from 100 to 180 MeV depending on the value of k^2. A matrix of 0.5 cm thick plastic scintillators in front of the array indicated whether the particle was a proton or a neutron and a 0.5 cm thick lead sheet placed in front of the array of 9 x 6 plastic counters (10 x 10 cm by 60 cm deep) reduced the low-energy background. The measurements were taken at $-k^2 = 0.2;\ 0.4;\ 0.6\ (\mathrm{GeV/c})^2$ and $\varepsilon = 0.98$.

The Daresbury-Pisa group working at NINA used a 2m x 2 m array of 145 plastic scintillation counters which allowed the determination of the production angles of the neutron. The momentum of the neutron was measured by time of flight techniques. This additional information allowed the measuring of the pion missing mass and thus improved the signal-to-noise ratio.

The array of veto counters placed in front of the neutron detector consisted of 26 pairs of scintillators separated by a sheet of iron 10 mm thick to absorb soft charged particles. The trajectories of the electrons were reconstructed from the data of six counter hodoscopes. The measurements were taken /151/ at $-k^2 = 0.078;\ 0.155;\ 0.233;\ 0.311\ (\mathrm{GeV/c})^2$ and $\varepsilon = 0.96$ and later /152/ at $-k^2 = 0.45;\ 0.58;\ 0.88\ (\mathrm{GeV/c})^2$ and $\varepsilon = 0.96$.

5.4.2 The Total Virtual Photon Cross Section

Assuming that only s- and p-wave multipoles contribute near threshold, the expansion in powers of q^* of the cross section can be cast into the form given by (5.17) with seven angular coefficients $a_0, a_1 \dots a_6$. The only coefficient required for the determination of $G_A(t)$ is a_0, the slope of the cross section at threshold, i.e.,

$$\frac{4\pi}{\Gamma_t} \lim_{q^* \to 0} \frac{k_L}{q^*} \frac{d^5\sigma}{dl_{02} d\Omega_1 d\Omega_\pi^*} = \lim_{q^* \to 0} (a_0 + bq^{*2}) = a_0;$$
$$b = \frac{1}{3} a_2 + a_5. \tag{5.28}$$

a_0 is the only quantity measured by the Frascati group as a consequence of the geometry adopted for the neutron detector. The other two groups determined five angular coefficients by a best fit of their data.

The determination of the a_0 coefficient is sensitive to the value of b used in the fit. The Frascati and DESY groups made the best fit of their data with a_0 and b as free parameters. The same procedure was used by the NINA group in its last paper /152/. In their previous work /151/ they introduced in the fits theoretical values of $b(k^2)$ chosen to cover the range of predictions by various models and included a systematic error on a_0 to account for the uncertainty in b. The two analysis procedures gave the same result.

Figure 5.11 shows the values of a_0 obtained by the three groups. The agreement is satisfactory. A correction, however, would be necessary because the values of the

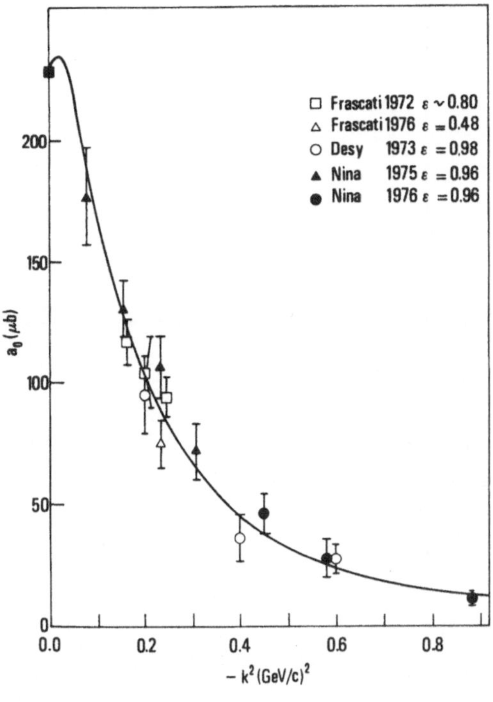

Fig. 5.11. Values of the a_0 coefficient [the threshold cross section slope: see 5.20)] versus $-k^2$. Note that the experimental points obtained by various authors refer to different values of the polarization parameter ε

parameter ε for the Frascati data (ε = 0.84, 0.74, 0.81 /150/ and ε = 0.48 /157/) are different from those of the other groups (ε = 0.98, DESY; ε = 0.96, NINA). The importance of such a correction is larger if σ_L at threshold is important (see Section 5.4.5). Under the assumption that σ_L is about equal to σ_T, the Frascati point would go up about 6, 15, and 10 %.

In Fig. 5.11 we also show the predictions of the fixed-t dispersion relation model of DEVENISH and LYTH /138/. The VON GEHLEN and WESSEL /67/ curve is about 20 μb higher for the whole range of k^2 values. Both these models incorporate, besides the standard pseudoscalar-coupling Born terms, a dispersion integral over the πN resonances. DEVENISH and LYTH include all higher resonances; VON GEHLEN and WESSEL consider only the Δ(1232). Both models use similar parametrizations for F_π and G_E^n. The value of the cross section at threshold is sensitive to G_E^n; the choice of negative rather than positive G_E^n gives predictions 50 % higher /67/. DEVENISH and LYTH also agree better with photoproduction for which, from a data compilation /153/ the NINA group estimates a_0 = 228 ± 8 μb. READ /154/ using only the data of ADAMOVICH et al. /155/ estimates a_0 = 234 ± 11 μb.

5.4.3 The Angular Coefficients

The angular coefficients $a_0 \ldots a_4$ in (5.18) have been determined by the DESY group as a function of W (varying from W = 1.079 to 1.12 GeV) for k^2 = 0.2 (GeV/c)^2 and by the NINA

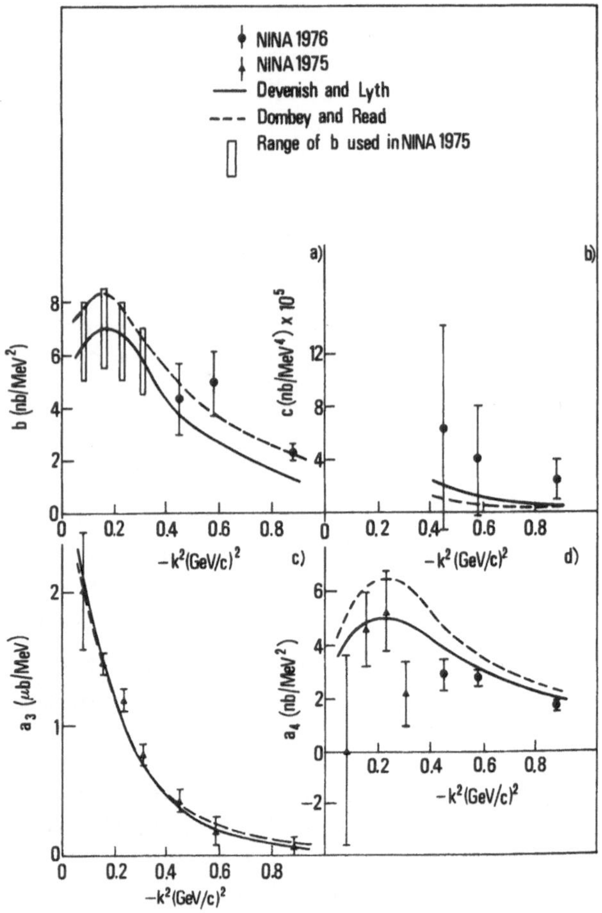

Fig. 5.12. Values of the other coefficients appearing in (5.18) versus k^2 obtained by the NINA group '76 /152/

group as a function of k^2 for values of W in the range W = 1.079 to 1.144 GeV. In an earlier paper /151/ the analysis was made with respect to both the polar and azimuthal angles θ_π^* and ϕ_π^*. In the second paper /152/ the analysis was only azimuthal. The values of the angular coefficients at threshold obtained from the last azimuthal analysis are compared in Fig. 5.12 with the predictions of the above-mentioned model as well as with those obtained by DOMBEY and READ /86/. The fairly good agreement with the experimental results confirms the adequacy of the description of p-waves in these models. In particular these authors found that the coefficient a_1 in (5.17) is well reproduced in those same theoretical models while the corresponding coefficient measured by the DESY group ($\bar{A}_1$ in their notation) is in serious disagreement.

5.4.4 The Separation of σ_L from σ_T at Threshold

The experimental problem of the separation of τ_L from σ_T near threshold is very interesting and at the same time rather difficult.

The determination of σ_L is of great physical interest because only virtual photons can induce longitudinal transitions. Furthermore, its value as a function of W and k^2 is more sensitive to the models adopted for describing the hadronic system than the electroproduction cross section or the transversal virtual photon cross section σ_T (W, k^2).

The best procedure for separating σ_L from σ_T consists in checking the ε and ϕ_π^* dependence of the coincidence cross section.

A less stringent way is based on the assumption that the single-arm cross section depends linearly on ε. Such an assumption is equivalent to asserting the adequacy of the one-photon exchange assumption. The results of single-arm experiments have been discussed in a previous section.

Only two coincidence experiments have been made with the aim of separating τ_L from σ_T near threshold.

A Saclay group /156/ measured the coincidences between the scattered electron and the π^+ produced in the direction of the virtual photons for

$$-k^2 = 1\ \text{fm}^{-2} = 0.038\ (\text{GeV/c})^2,\ W = 1175\ \text{GeV},$$

and two values of the polarization parameter (ε = 0.20 and 0.65). Magnetic spectrometers were used in both arms. The results are given in Table. 5.4.

The authors compare the values of σ_T with the predictions obtained from the isobaric model of Cochard, under two alternative assumptions: 1) $F_\pi = F_1^V$ (PS coupling); 2) $F_\pi = F_1^V$ (PV coupling). Both fit the data rather well, although the second seems to be slightly better.

Table 5.4. Electroproduction of π^+ at W = 1175 MeV. Separated cross section from the coincidence experiment of /156/

k^2 [fm^{-2}]	-1	-2	-3
$d\sigma_L/d\Omega_\pi$ [μb/sr]	7.8 ± 3.1	10.8 ± 2.2	13.1 ± 1.8
$d\sigma_T/d\Omega_\pi$ [μb/sr]	6.5 ± 1.0	5.2 ± 1.1	4.3 ± 0.7

From the values of σ_L, using the DOMBEY and READ model, they deduce estimates of $F_\pi(k^2)$. By parametrizing this quantity by means of the monopole expression (5.15) they deduce

$$\langle r_\pi^2\rangle^{1/2} = 0.74 \begin{matrix} +\ 0.11 \\ -\ 0.13 \end{matrix} \text{ fm.} \tag{5.29}$$

A Frascati group /157/ has attempted to measure the total cross section at $-k^2 = 6\ \text{fm}^{-2} \simeq 0.23\ (\text{GeV/c})^2$ with the same technique used at higher polarization values, i.e. by recording the coincidence between the inelastically scattered electron and the recoiling neutron. The measurements were taken in three intervals of values of W each of total width 6 %, equal to 1081, 1094, and 1111 MeV corresponding to the following central values of q^*: 24, 62, and 91 MeV/c.

By analysing these data together with those obtained at a higher polarization by the same group and the Daresbury-Pisa group, they find at threshold ($-k^2 = 6\ \text{fm}^{-2}$)

$$\sigma_T = 9.0 \pm 4{,}2\ \mu\text{b/sr},\quad \sigma_L = 11{,}5 \pm 5{,}6\ \mu\text{b/sr}, \tag{5.30}$$

which, although affected by a large error, seems to indicate the importance of the longitudinal cross section at threshold.

Separate determinations of the transversal and longitudinal parts have recently been reported /158/ for a single-arm experiment of threshold electroproduction on hydrogen. In this experimental configuration both π^+ and π^0 contribute to the measured cross sections so that a direct comparison with the previous results is not immediate (even if photoproduction and the experimental indications in electroproduction /2/ seem to suggest that the π^0 part represents a small contribution, $\lesssim 10$ %, to the complete matrix element). The reported values for σ_L/σ_T are

$$\sigma_L/\sigma_T = 0.31 \pm 0.18;\ 0.45 \pm 0.22;\ 0.51 \pm 0.20$$

at

$$k^2 = -7\ m_\pi^2,\ -10\ m_\pi^2,\ -16\ m_\pi^2.$$

5.4.5 Determination of $G_A(t)$

A) We now proceed to discuss the use of the pion electroproduction near threshold as an alternative source of information on the nucleon axial form factors. As already pointed out, extracting these quantities from experiments is not immediate and requires, for the physical process, a precise theoretical description, which embodies besides the chiral symmetry input also an indication on the form and size of the corrections arising from the nonvanishing pion mass.

Different approaches have been proposed and correspondingly there exist different theoretical formulae to be used for this specific interpretation of experimental electroproduction data. In our discussion of Section 3 we have mainly described a method based on the saturation of equal-time commutator matrix elements (we shall indicate it as the FPV model). As visualized in (3.71) the physical electroproduction amplitude is unambiguously expressed in terms of the electromagnetic and weak nucleon form factors $G_A(t)$ (the transversal part) and $D(t)$ (longitudinal part), while for the additional pieces an explicit recipe is provided. Although a complete evaluation of these terms is not always easy, they are of the order of m_π and often play, in particular for the transversal charged pion cross section, the role of corrections.

A different philosophy is, on the other hand, possible and it has been adopted by other authors. The main idea is to use for the electroproduction amplitude a representation based either on Feynman polar diagrams /86, 154/ with pseudovector pion-nucleon coupling (the D.R. model) or on dispersion relations /84/ (B.N.R. model). In so doing, among the other terms, the pion form factor $F_\pi(k^2)$ is introduced, either directly or to parametrize the high-energy tail of dispersive integrals. Then the requirements of gauge invariance and chiral invariance are enforced and used to fix the contact terms and subtraction constants, respectively. This procedure introduces in the formulae the form factor $G_A(t)$ which represents, together with $F_\pi(k^2)$,[26] the interesting parameters assuming the nucleon electromagnetic form factors to be known[27]. The same remarks apply to the improved soft-pion formulation of /83/ (N.Y. model).

Let us first examine the determinations of $G_A(k^2)$ which are obtained from double-arm experiments.

Table 5.5 shows the values of $G_A(k^2)/G_A(0)$ obtained by the NINA group using the various theories. In those calculations the standard dipole fits of Section 2.2.2. have been used for the electromagnetic nucleon form factors; furthermore, for $F_\pi(k^2)$ both D.R. and B.N.R. take

$$F_\pi(k^2) = F_1^V(k^2), \qquad (5.31)$$

while $F_\pi(k^2) = G_E^p(k^2)$ seems unlikely.

[26] It can be useful to remember that at threshold t and k^2 are linearly related, i.e.,

$$(1+m_\pi/m_N)(t)_{th.} = k^2 - m_\pi^2.$$

[27] Of course a comparison between the different models leads to a consistency relation between $D(t)$ and $F_\pi(k^2)$, which is, however, very hard to exploit owing to the crucial role of the correction terms.

Table 5.5. Values of $G_A(k^2)/G_A(0)$ deduced by the Daresbury-Pisa group

$-k^2$ $[GeV/c]^2$	Soft-pion limit	D.R.	B.N.R.	F.P.V.
0.078	0.815 ± 0.045	$0.873^{+0.059}_{-0.076}$	$0.795^{+0.063}_{-0.080}$	$0.782^{+0.046}_{-0.057}$
0.155	0.744 ± 0.031	$0.808^{+0.036}_{-0.041}$	$0.740^{+0.036}_{-0.041}$	0.722 ± 0.040
0.233	0.744 ± 0.039	$0.766^{+0.039}_{-0.044}$	$0.707^{+0.045}_{-0.050}$	$0.686^{+0.046}_{-0.048}$
0.311	0.650 ± 0.038	0.660 ± 0.045	$0.591^{+0.046}_{-0.058}$	$0.582^{+0.049}_{-.0.050}$
0.45	$0.538^{+0.034}_{-0.036}$	$0.562^{+0.039}_{-0.042}$	$0.495^{+0.045}_{-0.050}$	$0.463^{+0.050}_{-0.075}$
0.58	$0.450^{+0.043}_{-0.049}$	$0.462^{+0.050}_{-0.061}$	$0.390^{+0.055}_{-0.080}$	$0.385^{+0.048}_{-0.073}$
0.88	$0.317^{+0.025}_{-C.027}$	$0.321^{+0.025}_{-0.029}$	$0.245^{+0.040}_{-0.070}$	$0.272^{+0.044}_{-0.062}$

For D(t), F.P.V. adopt the simple pion dominated form

$$D(t) \simeq 2\, f_\pi g_{\pi N} \frac{m_\pi^2}{m_\pi^2 - t} \; . \tag{5.32}$$

It is customary to parametrize the axial vector form factor by means of either a dipole formula

$$G_A(k^2) = G_A(0)\,(1 - k^2/M_A^2)^{-2} \tag{5.33}$$

or a monopole formula

$$G_A(k^2) = G_A(0)\,(1 - k^2/M_A^2)^{-1} , \tag{5.34}$$

neither of which necessarily gives any physical significance to the mass parameter M_A. Both provide a convenient one-parameter representation of the data.

In Table 5.6 we present the values obtained for M_A by the three groups. The 1976 value is the result of a fit to all available data from threshold double-arm experiments.

From a glance at these results we notice that the D.R. model gives values of $G_A(k^2)/G_A(0)$ practically equal to the soft-pion theory (N.Y.) but somehow higher than F.P.V. or B.N.R. and consequently M_A turns out to be larger. Comparison with the values deduced from quasi-elastic neutrino scattering seems to favour the estimates for a lower M_A; indeed dipole fits to the differential cross section and to the total cross section give $M_A = 0.95 \pm 0.09$ GeV (see Section 2.2.3). The agreement is remarkable and confirms the substantial validity of the PCAC description, in spite of the slight disagreement between the theoretical models (see also Fig. 5.13).

These differences can be ascribed mainly to the theoretical uncertainties on the longitudinal cross section (whose prediction is actually beyond the domain of chiral symmetry). For this quantity the various models offer different estimates, already at the level of the polar terms, which contain $D(t)/m_\pi$ in F.P.V. and $F_\pi(k^2)/m_\pi^2 - t$ for D.R. and B.N.R. For instance, for $|k^2| \simeq 0.23\ (\mathrm{GeV}/c)^2$, which is the kinematical point of the Frascati experiment /157/ on σ_L/σ_T, the F.P.V. model predicts $R \simeq 0.4$, while lower values seem to be obtained by D.R. and B.N.R. ($R \simeq 0.1$ to 0.2). The experimental error is too large, however, to allow a precise test.

Separate experimental information on σ_L and σ_T at threshold would therefore be of great help in providing:

1) a neater determination of $G_A(t)$ from σ_T only, where the dependence on the theoretical model is small,

Table 5.6. Values for M_A from threshold electroproduction

Experiment	Monopole	Dipole	Model
Frascati (1972) /150/	0.68 ± 0.04	1.02 ± 0.04	B.N.R.
	0.70 ± 0.04	1.02 ± 0.04	F.P.V.
DESY (1973) /98/	0.68 ± 0.04	1.06 ± 0.06	D.R.
NINA (1975) /151/	0.77 ± 0.04(3)[a]	1.14 ± 0.06(3)	N.Y.
	0.76 ± 0.04(4)	1.12 ± 0.05(4)	N.Y.
	0.82 ± 0.05(3)	1.20 ± 0.07(3)	D.R.
	0.82 ± 0.05(4)	1.20 ± 0.07(4)	D.R.
	0.69 ± 0.05(3)	1.03 ± 0.06(3)	B.N.R.
	0.68 ± 0.05(4)	1.02 ± 0.06(4)	B.N.R.
	0.67 ± 0.04(3)	1.00 ± 0.06(3)	F.P.V.
	0.65 ± 0.04(4)	0.98 ± 0.06(4)	F.P.V.
NINA (1976) /152/	0.68 ± 0.03	1.08 ± 0.04	N.Y.
	0.69 ± 0.04	1.10 ± 0.04	D.R.
	0.62 ± 0.04	0.99 ± 0.06	B.N.R.
	/	0.96 ± 0.08	F.P.V.
Rome-Trieste /159/	0.70 ± 0.07	0.96 ± 0.19	Model-independent fit

[a] The number in parentheses indicates the number of data points used in the fit.

2) the possibility of an independent fit for the form factors appearing in σ_L according to the model or, conversely, of a test on the validity of the model itself.

One can finally notice that the measurements at higher $|k^2|$ allow the possibility of discriminating between the monopole and dipole fits. The Daresbury-Pisa group /152/ has investigated this point using the B.N.R. model (the F.P.V. model gives analogous results). They have fitted the function

$$\frac{G_A(k^2)}{G_A(0)} = (1 - 2\,C_A k^2 + C_B k^4)^{-1} \qquad (5.35)$$

and evaluated the best fit values of C_A, C_B. For $C_A = C_B$ the function (5.35) is simply the dipole, for $C_B = 0$ the monopole. The χ^2-contours of Fig. 5.14 shows a clear preference for the dipole formula although the monopole is not rejected at the level of 1.5 standard deviations.

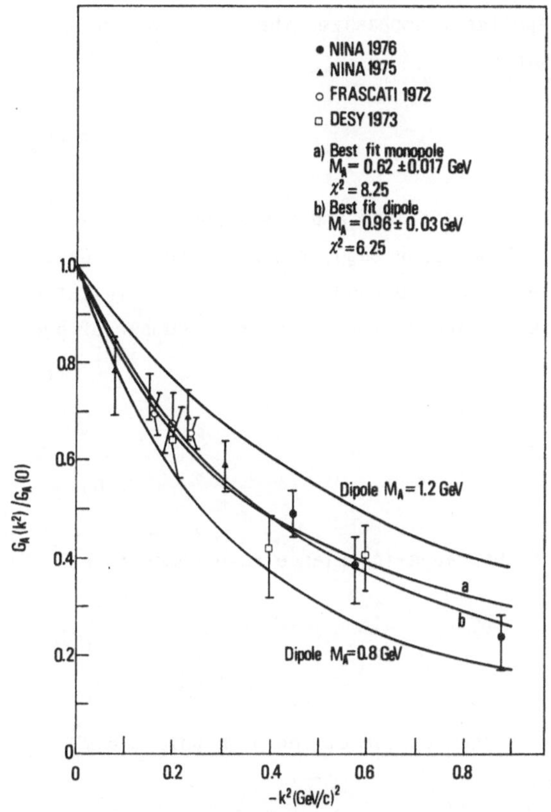

Fig. 5.13. Axial vector form factor $G_A(k^2)$ deduced using the Benfatto, Nicolò and Rossi model. The best fits to a dipole (b) and monopole (a) parametrization are shown

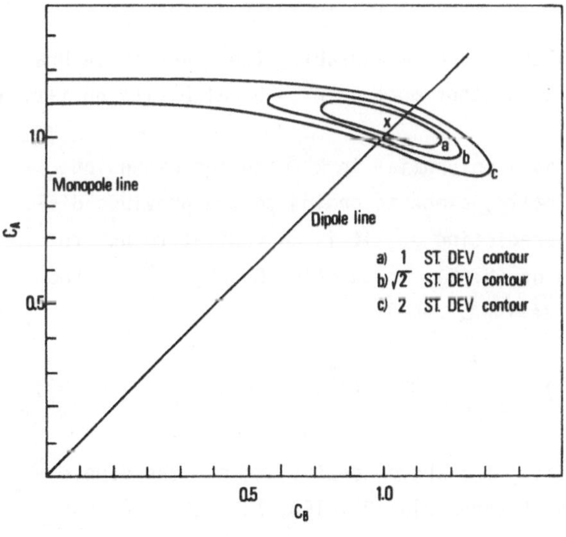

Fig. 5.14. χ^2-contours for the fit described in the text, showing the preference for a dipole over a monopole parametrization (NINA 1976: /152/)

B) In the previous discussions we have repeatedly emphasized the importance of a separate knowledge, beside σ_T, of the quantity

$$R(k^2) = (\sigma_L/\sigma_T)_{th.} \equiv -k^2 \left|\frac{L_{0+}}{k_0 E_{0+}}\right|^2 . \qquad (5.36)$$

Since the theoretical models offer different predictions and the experimental information is, at the moment, fairly poor, it has been suggested /159/ that a reasonable alternative may be represented by the use of a phenomenological parametrization of $R(k^2)$, for not too large $|k^2|$. The task is made easier thanks to a number of general constraints. These are

$$R = 0 \quad \text{at} \quad k^2 = 0, \qquad \text{a)}$$

$$R = 1 \quad \text{at} \quad k^2 = m_\pi^2. \qquad \text{b)}$$

[This second condition is a consequence of the gauge-invariance requirement; see Appendix (C.13)

$$E_{0^+}/L_{0^+} = 1 \quad \text{at} \quad \underline{k}_{CM} = \underline{0}].$$

One can further exploit the definition of the residue at the pion pole $t = m_\pi^2$, i.e., $k^2 = k_\pi^2 = m_\pi^2\ (2 + m_\pi/m_N)$ which gives

$$\lim_{k^2 \to k_\pi^2} (k^2-k_\pi^2)^2 R(k^2) = -k_\pi^2 \left[\frac{F_\pi(k_\pi^2) g_{\pi N}}{E_{0+}(k_\pi^2)}\right]^2 C; \qquad (5.37)$$

C is a known constant, $C = 1+O(m_\pi/m_N)$. Although not measurable, the quantity in brackets should not appreciably differ from the photoproduction $k^2 = 0$ value; its deviation is left as a free parameter.

$[(k^2-k_\pi^2)^2/k^2]R(k^2)$ is then expressed as a polynomial in k^2 with the expansion coefficients left as free parameters. Finally, since according to the previous discussion all models practically agree in predicting σ_T, it is convenient to use for E_{0+} a simple representation, of the kind of (3.62), in terms of $G_A(k^2)$, $G_M^V(k^2)$ and of an overall correction $\delta(k^2)$ to be determined, namely

$$(E_{0^+}(k^2))_{th.} = \sqrt{\frac{4m_N^2-k^2}{4m_N^2}} \cdot \frac{1}{\sqrt{2} f_\pi} \left| G_A(k^2) + \frac{k^2}{4m_N^2} G_A(0)\, G_M^V(k^2) + \delta(k^2) \right| . \qquad (5.38)$$

The free parameters can be adjusted using the above constraints and the double arm experiments for π^+ electroproduction at threshold /98, 150, 151, 152/, in particular the direct measurement of $R(k^2)$ in /157/.

Accepting the dipole fit this procedure gives for M_A the value

$$M_A = 0.96 \pm 0.19 \text{ GeV}. \tag{5.39}$$

The rather large error is the price one has to pay for the use of a nearly model independent fit and for the lack of enough experimental data.

C) The result (5.39) for M_A is compatible with the previous ones and with the neutrino data. The substantial, and remarkable, agreement among the various determinations of M_A indicates that time has perhaps come to turn the interpretation of electroproduction experiments: assuming $G_A(k^2)$ as given by neutrino scattering, one can predict σ_T and extract from the complete cross section the information on the longitudinal part and its theoretical content.

As the outcome of such a procedure one can obtain $(L_{0^+}/k_0)_{th.}$ as a function of k^2. This result can be used to determine, for π^+ electroproduction, $F_\pi(k^2)$ (in the D.R. and B.N.R. models) or D(t) (in the F.P.V. model). This is exemplified in Fig. 5.15 where one can see the phenomenological values of the longitudinal multipole and some relevant fits. $L_{0^+}(k^2)$ has been obtained from the experimental cross section assuming the validity for $E_{0^+}(k^2)$ of the expression (5.38) with $M_A = 0.96$ GeV and $\delta(k^2) = \text{const} \simeq 0.014\ m_\pi^{-1}$.

For a first interpretation of the π^+ data the following simple parametrization can be used /160/.

$$\frac{L_{0^+}(k^2)}{k_0} = \sqrt{\frac{4m_N^2-k^2}{4m_N^2}} \cdot \left\{ [D(t) - 2\, m_N\, G_A(0)] \cdot \frac{F_\pi(k_\pi^2)}{\sqrt{2}\, m_\pi f_\pi (2m_N+m_\pi)} \right\} + \frac{E_{0^+}(m_\pi^2)}{m_\pi} . \tag{5.40}$$

D(t) represents the equal time contribution while the other terms have been fixed by requiring consistency with the pion pole residue [see (5.37)] and the validity of the gauge-invariance constraint (C.13). For the longitudinal form factor D(t) it is fruitful to adopt the following expression

$$D(t) = \frac{2\, f_\pi g_{\pi N}\, m_\pi^2}{m_\pi^2 - t} + 2\, (m_N\, G_A(0) - f_\pi g_{\pi N}) \frac{\lambda^2}{\lambda^2 - t} , \tag{5.41}$$

which embodies the known constraints as $t \to 0$ and $m_\pi^2 \to 0$ plus an explicit contribution from the 3π cut, lumped, for simplicity, in a single polar term, whose position λ is a free parameter.

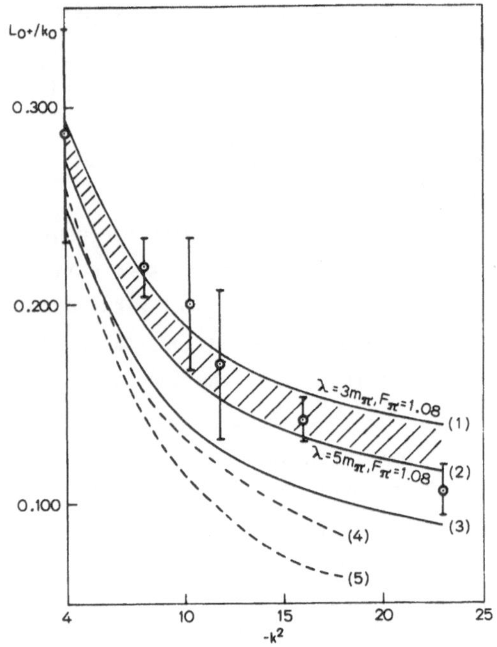

Fig. 5.15. The longitudinal π^+ multipole $(L_{0+}/k_0)_{th.}$ at threshold for some values of k^2 and the theoretical fits based on eqs. (5.40) and (5.41). These correspond to $\lambda=3m_\pi$, $F_\pi=1.08$ and 1.10 [curves (1) and (2)], $\lambda=5m_\pi$, $F_\pi=1.08$ and 1.10 [curves (2) and (3)], $\lambda=\infty$, $F_\pi=1.08$ and 1.10 [curves (4) and (5)]. L_{0+}/k_0 is in units m_π^{-2}, $-k^2$ in units m_π^2. (By accident the same curve (2) corresponds to two different fits.)

Taking for $F_\pi(k_\pi^2)$ the value given by ρ-dominance, the indication which is seen to emerge from Fig. 5.15 is that values of $\lambda = 3\ m_\pi \dots 5\ m_\pi$ lead to acceptable fits while higher values seem to be disfavoured. These results can be compared with the information on the pseudoscalar induced form factor $G_P(t)$ derived from μ-capture [see (2.40)]. One finds that for $\lambda = 3\ m_\pi \dots 5\ m_\pi$, $G_P(t = -0.88\ m_\pi^2) \simeq 8.8\ m_\mu^{-1}$, in the range of experimentally allowed values.

6. Other Developments

6.1 The Inverse Electroproduction Process

A) The reaction

$$\pi^-(q) + p(p_1) \to e^+(l_1) + e^-(l_2) + n(p_2) \tag{6.1}$$

has already been the object of investigation for the last two decades /161 - 164/. In the o.p.e.a. the electron pair is due to the materialization of a single time-like virtual photon and therefore the process (6.1) can be viewed as an inverse electroproduction (where, however, the photon is spacelike) (Fig. 6.1).

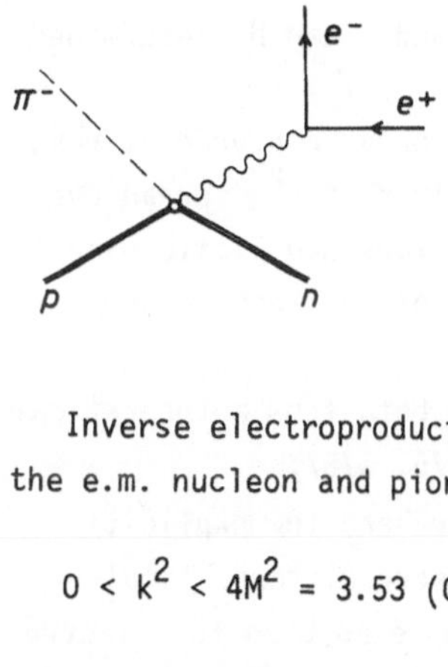

Fig. 6.1. Representation of the reaction $\pi^- p \to n e^+ e^-$ in the o.p.e.a. as "inverse electroproduction"

Inverse electroproduction is the only process which allows the determination of the e.m. nucleon and pion form factors in the intervals

$$0 < k^2 < 4M^2 = 3.53\ (\mathrm{GeV}/c)^2, \quad 0 < k^2 < 4\, m_\pi^2 \simeq 0.08\ (\mathrm{GeV}/c)^2,$$

which are kinematically unattainable from $e^+ e^-$ initial states.

Difficulties in the experimental study of this reaction arise from the need of a high rejection of competitive processes /165/: a) The cross section of $\pi^- p$ elastic scattering is $d\sigma/d\Omega \sim 10^{-27}$ cm^2/sr and is concentrated in the forward direction. Therefore the electrons and positrons of reaction (6.1) are conveniently detected at about 90^0 with respect to the π^--beam, where the elastically scattered hadrons are strongly reduced. b) The cross section for π^+ production, i.e.,

$$\pi^- + p \to n + \pi^- + \pi^+,$$

is about three orders of magnitude greater than that of reaction (6.1). The corresponding pions at 90° are very soft and can be suppressed strongly by threshold Cerenkov counters. c) The reactions

$$\pi^- + p \to \begin{cases} n + \pi^0 & \text{(a)} \\ p + \pi^0 + \pi^- & \text{(b)} \\ n + \pi^0 + \pi^0 & \text{(c)} \\ n + \gamma & \text{(d)} \\ p + \pi^- + \gamma & \text{(e)} \end{cases}$$

with a gamma ray converted into a Dalitz pair, contribute a rather unpleasant background. The most important processes are (a) and (d), which contribute about 60 % and 40 % of the counting rate due to capture in hydrogen of π^- at rest against 0.7 % from reaction (6.1).

Apart from the background problems, the events produced by π^- of well-defined momentum correspond to values of the virtual photon propagator k^2 spread from $(2m_l)^2$ to infinity with a frequency proportional to k^{-4}. Thus some of the advantages of electroproduction, i.e., the well-defined values of mass, energy, and polarization of all incident virtual photons, are lost, although k^2, k_0, and ε can be determined for each single event (Appendix A.4).

A few experimental studies of the process were made with the π^- captured at rest by protons /49, 166/. The events correspond to very small values of k^2, spread over a large relative interval. Therefore the hadron e.m. form factors had little effect on the process cross section, and only a rough determination of the mean square radius of the pion was possible from these measurements /49/.

More recently this process was studied by various authors both from a theoretical /167 - 174/ as well as an experimental point of view /165, 175, 176/.

B) Let us consider some theoretical aspects. We do not give here the explicit expression for the cross section nor discuss in detail the several theoretical descriptions of the phenomenon which, of course, parallel those adopted for electroproduction. It is, however, worth mentioning the interest of working in the particular kinematical configuration defined by the "pseudothreshold" condition /170, 171/

$$\underline{k}^* = \underline{l}_1 + \underline{l}_2 = \underline{0} \tag{6.2}$$

in the pion-nucleon centre-of-mass $\underline{p}_1 + \underline{q} = \underline{0}$[28], (i.e., $\underline{p}_2 = \underline{0}$).

At this point the cross section takes on a very simple form and, as a consequence of the behaviour of multipoles for small $|\underline{k}|$ and of gauge invariance, it can be expressed in terms of the E_{0+}, E_{2-} electric multipoles only /173/. One finds in particular, after integration over lepton variables,

$$\left.\frac{|\underline{q}|}{|\underline{k}|}\frac{d\sigma}{dk^2}\right)_{\underline{k}=\underline{0}} = \frac{\alpha^2}{3\pi}\,\frac{1}{k^2}\,\frac{M^2}{W^2}\left(2|H_1|^2 + |H_2|^2\right),$$

$$H_1 = E_{0+} + E_{2-}, \qquad (6.3)$$

$$H_2 = E_{0+} - 2E_{2-}.$$

At $\underline{k} = \underline{0}$, one actually has $k^2 = (W - M)^2$.

The most immediate advantage of working at pseudothreshold is that one can achieve a meaningful separation of the Born terms with respect to the continuum. Indeed only pion-nucleon resonances with $j^P = \frac{1}{2}^-, \frac{3}{2}^-$ can contribute[29]. Furthermore, for pions of kinetic energy between 100 and 360 MeV, i.e., $W \lesssim 1.5$ GeV, no resonances are excited (the lowest one is the $N^*(1520)$, $j^P = \frac{3}{2}^-$) so that the cross sections are expected to be determined mainly by the Born polar terms, which contain the nucleon and the pion electromagnetic form factors. As an indication we reproduce their explicit expressions in the c.m. system

$$\left.H_1\right)^{Born}_{\underline{k}=\underline{0}} = \sqrt{2}\, g_{\pi N}\sqrt{\frac{E_1+m_N}{2m_N}}\left\{\frac{F_1^n(k^2) - \sqrt{k^2}\,F_2^n(k^2)/2m_N}{\sqrt{k^2} + 2m_N} + \right.$$

$$\left.\frac{F_1^p(k^2) + \left[2(E_1-m_N) - \sqrt{k^2}\right]F_2^p(k^2)/2m_N}{\sqrt{k^2} - 2E_1}\right\} \qquad (6.4)$$

$$\left.H_2\right)^{Born}_{\underline{k}=\underline{0}} = \sqrt{2}\, g_{\pi N}\frac{|\underline{q}|^2}{2m_N}\sqrt{\frac{2m_N}{E_1+m_N}}\left\{\frac{F_\pi(k^2)}{k^2-2\omega_\pi\sqrt{k^2}} - \frac{F_1^p(k^2) - \sqrt{k^2}\,F_2^p(k^2)/2m_N}{k^2-2E_1\sqrt{k^2}}\right\} \qquad (6.5)$$

where

$$E_1 = W - \omega_\pi = \sqrt{k^2} + M - \omega_\pi, \qquad (6.6)$$

and all other symbols have the usual meaning.

---

[28] An additional prescription is actually required to fix the value of a scattering angle in that limit /173/.

[29] The final state is a s-wave with the nucleon and photon spins either parallel or antiparallel.

A quantitative estimate of the non-Born contributions has recently been given by BIETTI and PETRARCA /173/ on the basis of a simple model, and their conclusion is that, as long as $W \lesssim 1.5$ GeV, the corrections to the Born terms do not exceed 20 %.

An alternative description of inverse pion electroproduction, in particular at pseudothreshold, can be derived in the framework of current commutators and completeness /174/. Again, as long as $\sqrt{k^2}$ is of the order of a few pion masses, we expect that the equal-time commutator matrix elements plus the nucleon term represent the dominant contribution (with 10 to 20 % corrections from the continuum). Note, on the other hand, that $\underline{k} = \underline{0}$ is not a low-energy theorem point since the pion is in general moving. Only in the limit $k^2 \to m_\pi^2$, corresponding to $\underline{q} = \underline{0}$, i.e., to the full threshold configuration, we are in the presence of a genuine low-energy theorem, related to approximate chiral symmetry and with $O(m_\pi^2)$ corrections.

The interesting aspect of such a current algebra approach is that the approximate expressions obtained for the amplitudes H_1, H_2 by inserting the equal-time commutators and the nucleon terms show a very weak dependence on the nucleon electromagnetic form factors. In particular, retaining only the leading terms in k^2/m_N^2, t/m_N^2, one finds

$$H_1\Big)^{C.A.}_{\underline{k}=\underline{0}} \propto G_A(t) + \frac{4\bar{m}}{m_\pi^2}\sqrt{k^2}\, G_T^{(s)}(t), \tag{6.7}$$

$$H_2\Big)^{C.A.}_{\underline{k}=\underline{0}} \propto \frac{\sqrt{k^2}}{m_\pi^2}\left\{[D(t) - (1+\frac{\sqrt{k^2}}{2m_N})\, D(m_\pi^2-k^2)] + \frac{m_\pi^2}{2m_N}\,[G_A(m_\pi^2-k^2) - \frac{t}{2m_N}\, G_P(m_\pi^2-k^2)]\right\}, \tag{6.8}$$

where

$$t = (m_\pi^2-k^2)\,\frac{(1+\sqrt{k^2}/2m_N)^2}{(1+\sqrt{k^2}/m_N)},$$

and all other symbols have already been defined in the previous sections.

The expressions (6.4, 5) and (6.7, 8) exhibit simple, complementary descriptions to pseudothreshold inverse electroproduction in terms of electromagnetic form factors and of weak form factors, respectively. The possibility of new information on these quantities by fitting experimental data is not excluded and should be kept in mind, in spite of the objective difficulty of the experiments.

C) In all the most recent experiments due to Russian authors /175/ a well-collimated beam of π^- of 275 MeV kinetic energy (W = 1295 MeV) enters a hydrogen target, behind which a veto counter allows the selection of events in which the pion is absorbed.

The direction and energy of electrons and positrons emitted in each event are measured by two telescopes placed at opposite sides of the target. Each of these telescopes covers an angle interval $90^{0} \pm 20^{0}$ with respect to the beam and includes scintillation counters, a spark chamber, a threshold water Cerenkov counter (to suppress the background of low-energy pions) and a total absorption Cerenkov counter (to measure the energy of the electrons and to suppress the background).

The data for electrons and positrons of energies > 60 MeV emitted around 90^{0} allow a determination of the differential cross section for production of electron pairs. The cross section, integrated with respect to the electron energies is found (from 1100 events) to amount to /176/

$$\frac{d\sigma}{d\Omega_1 d\Omega_2} = (6.90 \pm 0.69) \cdot 10^{-33}\ cm^2/sr^2.$$

A sample of 234 events, divided into three intervals of values of k^2, was analysed assuming the validity of the approximate relation $F_1^V(k^2) = F_\pi(k^2)$ and neglecting $\mathrm{Im}\ F_1(k^2)$.

Under these simplifying assumptions the following values of $F_\pi(k^2)$ were obtained:

k^2/m_π^2	3.4	4.4	5.8	
$F_\pi(k^2)$	1.10 ± 0.07	1.14 ± 0.07	1.30 ± 0.07 ;	(6.9)
χ^2/d.f.	1.10/3	3.6/4	3.0/4	

These are plotted in Fig. 6.2 which shows, on a larger scale, the central part of Fig. 6.3. Although the experimental errors are appreciable, the continuity of $F_\pi(k^2)$ when k^2 passes from negative to positive values is certainly a striking feature of these figures.

The authors also try to analyse the same data assuming $F_\pi(k^2)$ and $F_1^V(k^2)$ as independent from each other (keeping $\mathrm{Im}\ F_\pi = \mathrm{Im}\ F_1^V = 0$), but in this case the relative errors turn out to be about 100 % for the two lower values of k^2 while for $k^2=5.8\ m_\pi^2$ they obtain

$$F_1^V(5.8\ m_\pi^2) = 1.4\ ^{+\ 0.1}_{-\ 0.2}, \quad F_\pi(5.8\ m_\pi^2) = 1.2\ ^{+\ 0.2}_{-\ 0.3}. \qquad (6.10)$$

Figure 6.3 summarizes our present experimental information on the charged pion electromagnetic form factor $F_\pi(k^2)$. The experimental points in the spacelike region have been discussed in Section 5.3 and are only a part of those shown in Fig. 5.3. The points in the timelike region obtained from $e^+e^- \to \pi^+\pi^-$ /31 - 34/ have already been mentioned briefly at the beginning of Section 2.2.1, while those for very small positive values of k^2 are the main result of this section.

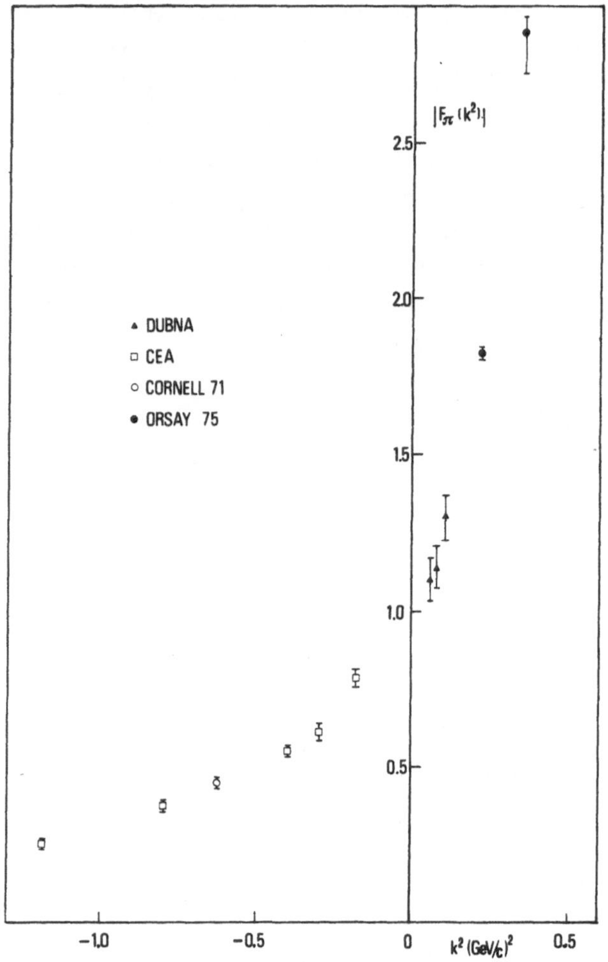

Fig. 6.2. Pion form factor obtained at Dubna from inverse electroproduction. Also points obtained from π^+ electroproduction ($k^2<0$) and $e^+e^- \to \pi^+\pi^-$ ($k^2 > 4m_\pi^2$) are plotted in this figure which shows, on an enlarged scale, the central part of Fig. 6.3.

The curve is a p-wave Breit-Wigner as suggested by GOUNARIS and SAKURAI /177/. It fits very satisfactorily the data in the vicinity of the peak due to the ρ-vector resonance, altered by its interference with the ω. The deviation of the experimental points for $k^2 > 1$ (GeV/c)^2 seems to indicate a contribution to the pion form factor originating from a higher vector meson (ρ', $M_{\rho'} \simeq 1250$ MeV).

The striking and gratifying feature of this picture is the overall consistency of the experimental determinations of $F_\pi(k^2)$ extracted from three quite different phenomena.

D) TKEBUCHAVA /219/ has obtained the values of the induced pseudoscalar form factor in the region $-t \lesssim 5\ m_\pi^2$ from an analysis of the inverse electroproduction reaction in the current algebra framework using (6.4, 5, 8) and the experimental values of $F_\pi(k^2)$ and $F_1^p(k^2)$ deduced in Dubna using the compensation properties of the non-Born

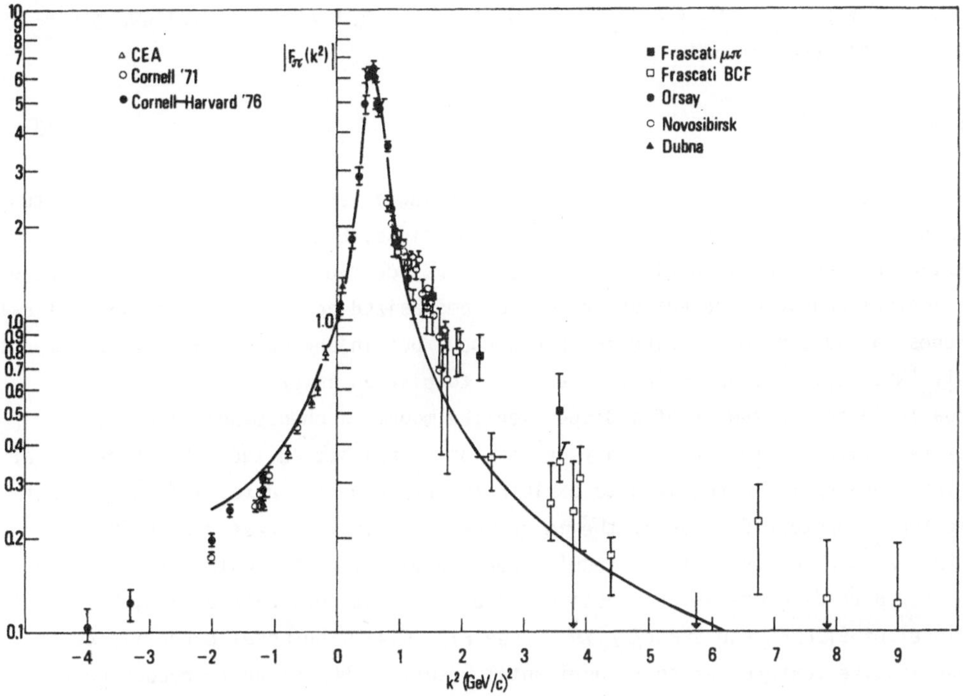

Fig. 6.3. Experimental results available at present on the electromagnetic pion form factor. The data for $k^2 < 0$ have been discussed in Section 5.2, those for $k^2 \geq 4m_\pi^2$ have been mentioned in Section 2.2.1, except those for $k^2 \gtrsim 0$ (Fig. 6.2). The continuous curve is given by the Gounaris-Sakurai model

amplitude /122/. It turns out that in the framework of the current algebra techniques, the inverse electroproduction reaction seems to be well suited for the determination of the induced nucleon pseudoscalar form factor $G_P(t)$. His analysis agrees with the indicative determination of $G_P(t)$ from direct electroproduction discussed in Section 5.4.5 C).

6.2 Electroproduction of the πΔ State Near Threshold

A) This section is devoted to a brief discussion of the two-pion electroproduction process

$$e + N \rightarrow e' + N' + \pi + \pi \qquad (6.11)$$

in the vicinity of $W_{th} = M_\Delta + m_\pi \simeq 1372$ MeV where, on hydrogen, the following production channels are expected to be important:

$$\gamma_V p \to \pi^+\Delta^0,\ \pi^-\Delta^{++}. \tag{6.12}$$

Photo- and electroproduction of the $\pi\Delta$ state provide a very good opportunity to study experimentally several features of the N-Δ transitions, exactly in the same way as information on the elastic nucleon form factors was derived from single-pion electroproduction. Of course different aspect can be emphasized according to the theoretical approaches one adopts to describe the phenomenon, but in one way or another we expect the $\langle\Delta|V_\mu|N\rangle$, $\langle\Delta|V_\mu|\Delta\rangle$, and $\langle\Delta|A_\mu|N\rangle$ vertices to play an interesting role.

Actually in the framework of a dispersionlike model, a phenomenological ($\gamma N\pi\Delta$) contact term (i.e., a term which only depends on k^2 and not on the other kinematical variables) appears to be required to explain the experimental data. A large contact interaction in photoproduction of the $\pi\Delta$ channel was first suggested by CUTKOSKY and ZACHARIASEN /178/ in a static model theory. Similarly, the contact term is one of the four Born diagrams which constitute the covariant, gauge-invariant,[30] electric polar model of STICHEL and SCHOLZ /179/, (see Fig. 6.4). The model correctly describes the quantitative features of the experimental results /180/ of photoproduction near threshold, such as isotropic production, rapid rise of the cross section, and its magnitude, only because of the presence of the contact term, which dominates at low $W \approx W_{th}$. In other words, in that region the contribution of s-channel resonant states alone is not enough to explain the experimental situation and the reaction is dominated by a real, nonresonant s-wave amplitude.

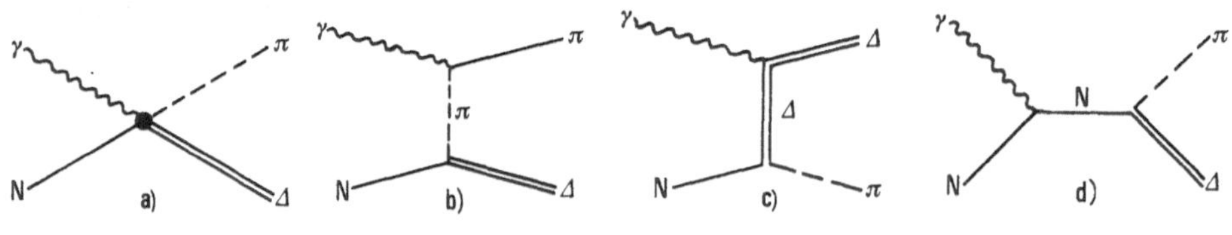

Fig. 6.4. Born terms used to describe $\gamma_V p \to \pi^-\Delta^{++}$ in the model of Stichel and Scholz

[30] Indeed the contact term can simply be motivated by the gauge-invariant requirement $\partial_\mu \to \partial_\mu - ie\mathscr{A}_\mu$ applied to the derivative ($N\pi\Delta$) coupling $(g^*/m_N)\bar\psi\psi_\mu\partial^\mu\pi$. Furthermore, only the orbital (spinless) part of the electromagnetic vertices is used.

The generalization to electroproduction, $k^2 \neq 0$, of the pole model of Fig. 6.4 is straightforward /181/. In the matrix element now enter the electric form factors of the particles and of the contact term, and the simplest choice is $F_{contact}(k^2) = F_\pi(k^2) = F_\Delta(k^2) = F_1^V(k^2)$. In this spirit BARTL et al. /182/ predict the longitudinal and transverse parts of the cross section starting from real photoproduction, in the framework of vector-meson dominance, under the assumption of approximate mass independence for the invariant amplitudes of the VN → πΔ process. In their approach the cross section turns out to be multiplied by the common form factor $(1-k^2/M_\rho^2)^{-1}$, which again allows a quantitative description of the threshold region. The model, however, has difficulties with the cross section of the channel $\gamma_V p \to \pi^+\Delta^0$, in particular the experimental value /183/

$$R_{exp} \frac{(\pi^-\Delta^{++})}{(\pi^+\Delta^0)} = 1.50 \pm 0.2, \tag{6.13}$$

observed at W = 2.23 GeV, $0.2 < -k^2 < 0.8$, $t - t_{min} = 0.04$ $(GeV/c)^2$ is lower than the predicted one ($R_{th} \approx 2.7$). Also the longitudinal parts seems to come out too low; the reader can find in the original paper a discussion on these points.

A complementary theoretical description of the behaviour around threshold is obtained by resorting to current algebra equal time commutators and to approximate chiral symmetry /184/. This approach leads to an expression of the πΔ electroproduction amplitude in terms of the axial matrix element $\langle\Delta|A_\mu|N\rangle$, which plays the role of the analogue of the contact term[31], and of the electromagnetic vertices between the nucleon, the Δ and higher resonances. In specific theoretical models (based substantially on the Fermi-Watson theorem) the vertex $\langle\Delta|A_\mu|N\rangle$ is directly proportional to the nucleon axial form factor, thus suggesting the possibility of using the particular reaction $\gamma_V p \to \Delta^{++}\pi^-$ (soft) as an independent source of information on $G_A(k^2)$ /185/.

More recently /186/ the approach described in Section 3.5 for single pion electroproduction, and based on the saturation of equal time commutators in a suitable reference frame, has been adapted to provide a simple representation for the $\gamma_V p \to \Delta^{++}\pi^-$ process at threshold. It turns out that the $\langle\Delta^{++}|A_\mu|p\rangle$ vertex represents the most important part of the threshold amplitude, while the dependence on the quantity $\langle\Delta^{++}|V_\mu|p\rangle$ is practically negligible and, among the higher states, only the $D_{13}(1520)$

---

[31] This contribution has indeed been shown /185/ to coincide (in the static limit) in the case of a real photon $k^2 = 0$ with the direct interaction term of the Cutkosky-Zachariasen model. Note that in principle the form factors in $\langle\Delta|A_\mu|N\rangle$ bear a dependence on t, but in the soft-pion limit $t \to k^2$.

resonance seems to give a sizeable effect. Comparison with the available experimental data allows then a determination of the form factors H_1, H_2, H_3 defined in (2.28, 29) 29). We discuss again these points in the next section.

B) The experimental study has been made by a few groups working at DESY /183, 187, 188/ and NINA /189/, which have investigated the $\pi^-\Delta^{++}$ channel, the $\pi^+\Delta^0$ channel, or both. Three of them have used two spectrometers for detecting in coincidence the scattered electron and the produced pion. In the fourth experiment /189/ a 7.2 GeV electron beam enters a streamer chamber and strikes a liquid hydrogen target placed inside it. A counter hodoscope of scintillation and shower counters and proportional chambers detect the scattered electron and trigger the chamber. The hadrons produced are detected over the full solid angle. The authors analyse 4000 events of the $ep \to e'p'\pi^+\pi^-$ type with $0.3 \leq -k^2 \leq 1.5$ $(GeV/c)^2$ and $1.3 < W < 2.8$ GeV. They construct the $(p\pi^+)(p\pi^-)$ and $(\pi^+\pi^-)$ mass distributions and determine the $\pi^-\Delta^{++}$ and $\pi^+\Delta^0$ cross section by a maximum likelihood fit to the Dalitz plot density

$$dN\,(M^2_{p\pi},\ M^2_{\pi\pi}) = [a_{\Delta^{++}}\ f_{\Delta^{++}}\ (M_{p\pi^+}) + a_{\Delta^0}\ f_{\Delta^0}\ (M_{p\pi^-}) + a_{ps}\ f_{ps}]\ dM^2_{p\pi}\ dM^2_{\pi^+\pi^-}, \quad (6.14)$$

where the a's are fit parameters which measure the individual contributions, $f_{\Delta^{++}}$ and f_{Δ^0} represent the corresponding Breit-Wigner terms, and f_{ps} a phase-space-like background.

The total cross section, averaged over k^2 [$\langle -k^2\rangle = 0.6$ $(GeV/c)^2$] goes through a maximum at W = 1.5 GeV and decreases rapidly at higher energies (Fig. 6.5) with an energy dependence similar to that found in photoproduction. The angular distribution in the threshold region is consistent with isotropic production, i.e., with dominant s-wave production. The same conclusion seems to be confirmed by the approximately linear rise of the cross section above threshold. A study of the Δ^{++} decay angular distribution shows that the Δ^{++} has predominantly helicity $\lambda_\Delta = 3/2$. The $(\pi^-\Delta^{++})$

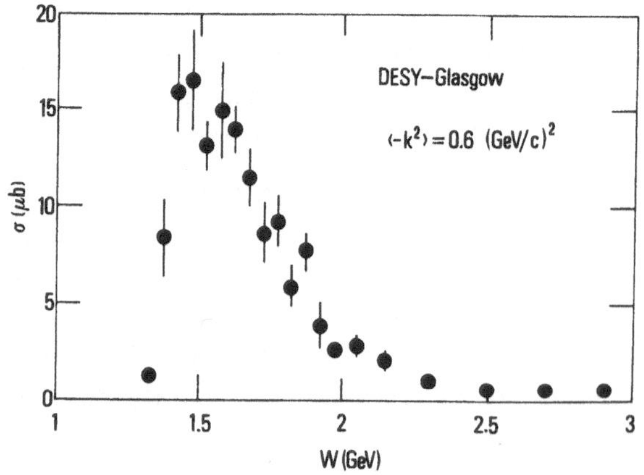

Fig. 6.5. The total cross section for $\gamma_V p \to \pi^-\Delta^{++}$ as a function of W averaged over the $-k^2$ interval 0.3 - 1.4 $(GeV/c)^2$

system is therefore in a $j^P = 3/2^-$ state. This excludes the possibility that P_{11} formation is responsible for $(\pi^-\Delta^{++})$ production near threshold. The observed production features are thus consistent with the dominance of the contact term as in the case of photoproduction.

Figure 6.6 shows the cross section dependence on k^2 where the decrease is well reproduced by the ρ-meson propagator $(1 - k^2/M_\rho^2)^{-2}$.

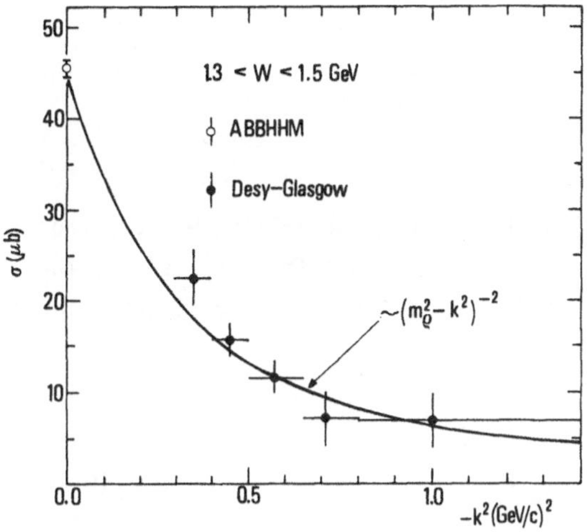

Fig. 6.6. The total cross section for $\gamma_v p \to \pi^-\Delta^{++}$ as a function of $-k^2$ in the threshold region (1.3 < W < 1.5 GeV) /189/

Finally the DESY-Glasgow group /190/ has also used its data to determine the axial transition $N\Delta$ form factors $H_i(k^2)$. Actually, in the framework of the theoretical model by ADLER and WEISBERGER already mentioned /185/, the cross section for $\pi^-\Delta^{++}$ electroproduction turns out to be very closely proportional to $G_A(k^2)$, as a consequence of the strong dominance of the current algebra equal time commutator. The result is shown in Fig. 6.7. A dipole fit parametrization, which include also single pion electroproduction data, gives

$$M_A = (1.16 \pm 0.03) \text{ GeV}, \tag{6.15}$$

a slightly higher value than the one obtained by single-pion electroproduction or in neutrino scattering.

A more direct determination of the form factors $H_i(k^2)$ has been recently presented in /186/. Using the experimental information on both photo- and electroproduction of $\pi\Delta$ at threshold (cross sections and spin density matrix elements) and after taking the finite width of the Δ resonance into account, a satisfactory agreement between theory and experiment can be obtained with the simple parametrization

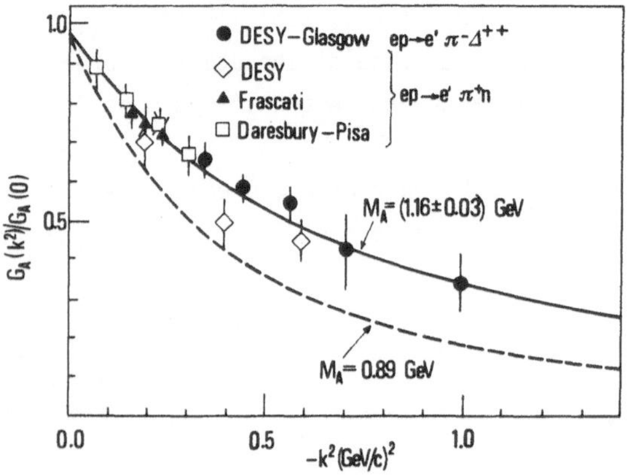

Fig. 6.7. The axial vector form factor $G_A(k^2)$ deduced by a DESY-Glasgow group from $ep \to e'\pi^-\Delta^{++}$ appears to agree fairly well with $G_A(k^2)$ deduced from $ep \to e'\pi^+n$. The only difference is that the value of M_A of an overall fit seems to be slightly larger (see Fig. 5.13)

$$H_i(k^2) = H_i(0)\ (1 - k^2/M_{A^*}^2)^{-2},\ i = 1, 2, 3. \tag{6.16}$$

The values of $H_i(0)$ are consistent with those given in (2.42) and derived from neutrino scattering, while for M_A^* one finds

$$M_A^* \simeq 1.1 \text{ GeV}. \tag{6.17}$$

The slight discrepancy with the neutrino scattering result, $M_A^* \simeq 0.96$ GeV, see (2.43), is presumably due to the different parametrizations adopted for the k^2 dependence of the form factors (actually the model of /186/ contains one more phenomenological parameter to be fitted, as is also the case of the $\nu N \to \mu\Delta$ process).

6.3 Electroproduction of η and k Mesons

The investigation of other mesons belonging to the same SU(3) octet as the pion is very interesting since one expects that models or theories of π-production may be applied to these other cases with only modifications of detail rather than substance. In the following our considerations will be limited to a few indicative points.

A) The Electroproduction of η

The reaction

$$e + p \to e' + p' + \eta \tag{6.18}$$

has been investigated by various authors /191 - 193/ at the resonance $S_{11}(1535)$, which has a large branching ratio to the decay channel $(p\eta)$. Indeed analysis of photoproduction data shows that more than 90 % of the η's from threshold up to W = 1.55 GeV come from the decay of the S_{11} resonance /194/. The physical problem bears considerable similarity to the electroproduction of π^0 at the P_{33} (1232) resonance.

In DESY's experiment /193/ the scattered electron is detected in a double focusing and vertically bending spectrometer and is identified by a CO_2 Cerenkov and sandwich shower counters. The recoiling proton is detected in coincidence in a nonfocusing spectrometer including two hodoscopes.

Protons are distinguished from π^+ meson by time of flight. The $(ep\eta)$ channel clearly shows up in missing mass spectra.

The data at $-k^2 = 0.6$ and 1 $(GeV/c)^2$ have been fitted to the angular dependence

$$\frac{d\sigma}{d\Omega^*} = A_0 + \varepsilon B_0 + (A_1 + \varepsilon B_1)\cos\theta^* + D_0\sqrt{2\varepsilon(\varepsilon+1)}\,\sin\theta^*\cos(2\phi^*), \tag{6.19}$$

based on the assumption that in the final state only contributions of s-wave, interference of s- and p-wave with total angular momentum $\frac{1}{2}$ are present. The scalar-transverse interference term D_0 is consistent with zero; up to W = 1625 the obtained standard deviations of D_0 are about 10 % of $A_0 + \varepsilon B_0$, which is the dominant term. Figure 6.8 shows the two terms

$$\sigma_{tot} = 4\pi\,(A_0 + \varepsilon B_0), \quad 4\pi\,(A_1 + \varepsilon B_1), \tag{6.20}$$

as a function of W. The total cross section can be represented with the Breit-Wigner expression

$$\sigma_{tot} = \frac{|\underline{q}_\eta^*|}{|\underline{k}^*|}\;\frac{A}{(W-W_{res})^2+\Gamma^2(W)/4}\;. \tag{6.21}$$

The width has been parametrized according to the branching ratio of the decay modes of S_{11} (1535)

$$\Gamma(W) = \Gamma_0\left(0.55\,\frac{|\underline{q}_\eta^*|}{|\underline{q}_\eta^*|_{res}} + 0.35\,\frac{|\underline{q}_\pi^*|}{|\underline{q}_\pi^*|_{res}} + 0.1\right). \tag{6.22}$$

The solid curves in Fig. 6.8 are least square fits of the above formula to the measured cross section with A, Γ_0 and W_{res} as free parameters.

The authors obtain

$-k^2$[GeV/c]^2	A[μb x GeV^2]	Γ_0[MeV]	W_{res} [MeV]
0.23	0.173	-	-
0.6	0.215	154	1526
1	0.204	147	1524

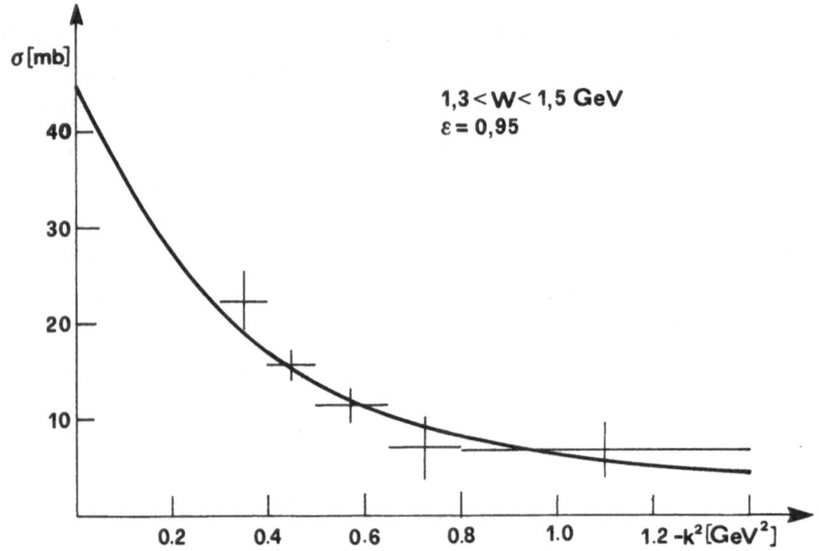

Fig. 6.8. Electroproduction of η: coefficients of angular distributions as functions of W for $-k^2 = 0.22$, 0.6 and 1.0 $(GeV/c)^2$. At $-k^2 = 0.22$ $(GeV/c)^2$ the coefficient $A_1 + \varepsilon B_1$ has been fixed at zero. Solid line: Breit-Wigner curve /193/

The DESY group interprets the term $A_0 + \varepsilon B_0$ as a hint of the presence of the resonance S_{11} (1535). Its strength at $-k^2 = 1$ $(GeV/c)^2$ is remarkable. The total production cross section at this point is only about 20 % smaller than the photoproduction cross section. Assuming a partial decay rate of 55 % for the $S_{11} \rightarrow p\eta$ channel, the resonance S_{11} (1535) contributes, at W = 1535, $-k^2 = 1$ $(GeV/c)^2$, about 19 μb. This is more than 20 % of the total $\gamma_V p$ cross section of about 80 μb. At $k^2 = 0$ the resonance S_{11} (1535) contributes at most 10 % to the total cross section.

Since the second bump in the total $\gamma_V p$ cross section remains equally prominent between $k^2 = 0$ and $-k^2 = 1$ $(GeV/c)^2$, the results of the DESY group imply that the resonance D_{13} (1520) decreases faster with increasing momentum transfer than the total $\gamma_V p$ cross section.

The interesting problem that cannot be decided by the experimental data available today is whether the large cross section of S_{11} (1535) at spacelike momentum transfer is due to scalar excitation.

Figure 6.9 shows the values of σ_{tot} at the resonance (W = 1535 MeV) obtained by the three groups /191 - 193/ as a function of k^2. The theoretical curves correspond to different versions /195, 196/ of a relativistic quark model for electroproduction of resonances.

The difference between the values obtained at DESY and those by the other two groups may be due to the strong W dependence of the cross section and to the fact that the results of /191, 192/ refer to an average over a wider range of W-values.

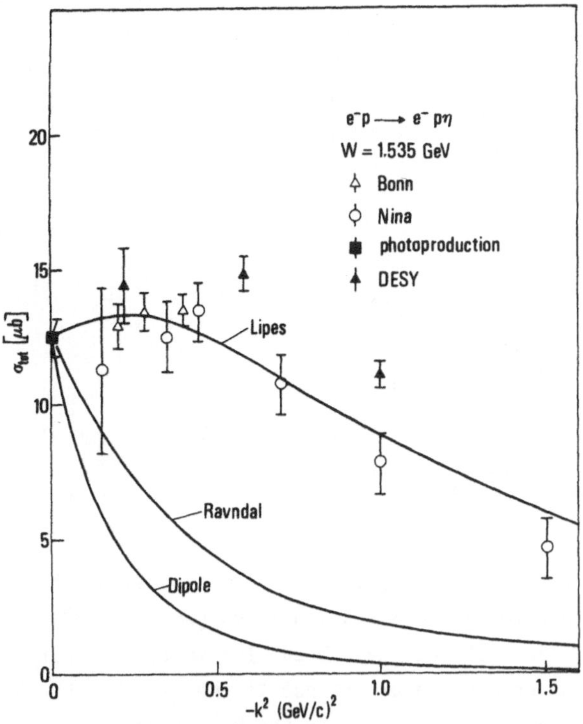

Fig. 6.9. Total cross section for $\gamma_v p \to \eta p$ at W=1.535 GeV as a function of $-k^2$. Solid lines: quark model calculations by RAVNDAL /195/ and LIPES /196/ and the dipole form factor, all normalized to k^2=0

B) Electroproduction of K

The reactions

$$e + p \to e' + K^+ + \Lambda \qquad (W_{th} = 1609 \text{ MeV}), \tag{6.23}$$

$$e + p \to e' + K^+ + \Sigma^0 \qquad (W_{th} = 1689 \text{ MeV}), \tag{6.24}$$

have been investigated by CEA /197/, DESY /198/ and Cornell-Harvard groups /199/. Two magnetic spectrometers were used to detect the scattered electron and the electroproduced kaon. Electrons were identified by combining the outputs of Cerenkov counters and shower counters of various types. Time of flight and threshold Cerenkov counters were used to separate pions, kaons, and protons. Besides that the usual corrections due to the decay in flight of K^+ should be taken into account (20 to 100 % depending on $\underline{q}_k$).

Figure 6.10 shows the K^+ cross section as a function of the invariant mass W at $\langle -k^2 \rangle = 0.29$ (GeV/c)^2, $\langle\theta\rangle = 6^0$, and $\langle\varepsilon\rangle = 0.86$ /197/. The dashed curve shows the energy dependence of a cross section proportional to W^{-2}, which was plotted for comparison. In single K^+ photoproduction at large W, the invariant matrix element is energy independent /200/. If the matrix element shows the same property also in single K^+ electroproduction, the corresponding cross section $d\sigma/d\Omega$ should decrease at large values of W, as W^{-2}.

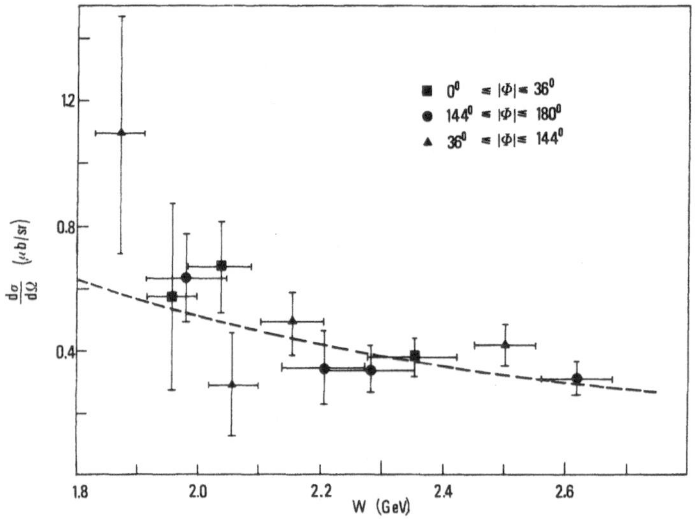

Fig. 6.10. The W-dependence of the $K^+\Lambda$ virtual photoproduction cross section. The dashed curve shows the energy dependence for a cross section which goes at $1/W^2$. For these data $\langle -k^2 \rangle \simeq 0.29$ $(\mathrm{GeV/c})^2$, $\langle \theta \rangle \simeq 6^0$ and $\langle \varepsilon \rangle \simeq 0.86$. The vertical error bars take into account only statistical errors and errors due to the fit. The horizontal error bars give the rms variation of the energy over the bins /197/

The results of the Cornell-Harvard-group /199/ confirm and extend those of the other groups, i.e.,

a) the ratio of the cross sections $K^+\Sigma^0/K^+\Lambda$ is not larger than 0.25;

b) the cross section $K^+\Sigma^0$ decreases rapidly with k^2, while in the $K^+\Lambda$ case the k^2 dependence is weak (Fig. 6.11).

Photoproduction measurements made in the same momentum region but higher energies /200/ and lower energy but higher momentum transfer give Σ^0/Λ ratios typically between 0.5 and 1.

Point b) is in agreement with what is naively expected: since the coupling constant $G^2_{Nk\Sigma}$ is supposed to be smaller than $G^2_{Nk\Lambda}$ /201/, K exchange will play a minor role in $K^+\Sigma^0$ production but will contribute significantly to the $K^+\Lambda$ cross section, in particular to its longitudinal part. A quantitative description of both photo- and electroproduction data for $K^+\Sigma^0$, and $K^+\Lambda$ requires K, K^* (892), K^* (1470) and either K_A (1^{+-}) or K_B (1^{++}) exchange /202/.

No data are available in the W region near threshold. Conversely, no systematic current algebra predictions exist for the electroproduction of K mesons. (Among the few calculations available for K-photoproduction we quote the one in /203/, where earlier references can be found.) This perhaps reflects the feeling that for strange axial vector currents the breaking of SU(3) x SU(3) chiral symmetry is related to the K meson mass, i.e., $\partial^\mu A_\mu^{(k)} \propto m_K^2$ (generalized PCAC), and one expects large deviations

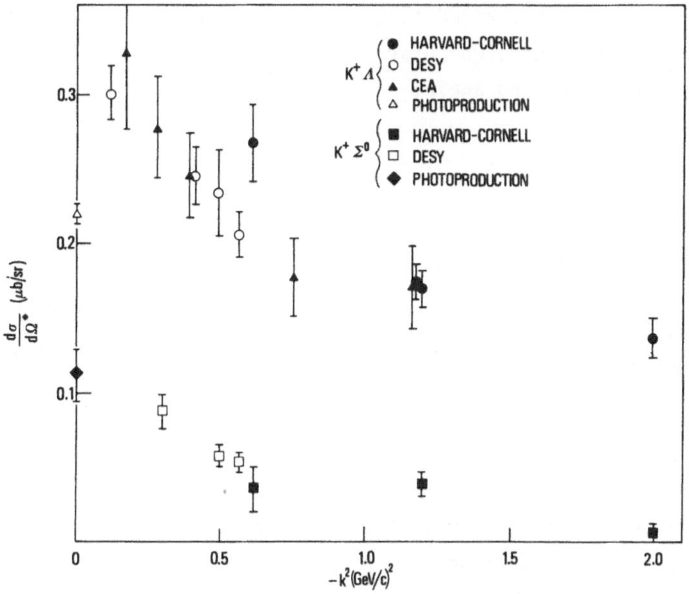

Fig. 6.11. k^2 dependence for the $K^+\Lambda$ and $K^+\Sigma^0$ cross sections at W = 2.66 GeV. Also shown are photoproduction measurements /199/

from the soft K theorem. Furthermore, besides the equal time commutator matrix element $\langle\Lambda|A_\mu|N\rangle$, there is the contribution of the s-channel intermediate states, whose structure is more complicated now, and which should be carefully taken into account. Anyway a systematic analysis, both theoretical and experimental, of strange particle photo- and electroproduction near threshold would represent a not trivial test of the existence of an approximate SU(3) x SU(3) symmetry in nature and of its limits of validity. A similar approach to meson-baryon scattering lengths has led to interesting indications on the structure and properties of the quark currents /204/.

6.4 Electroproduction with Polarized Beams and Polarized Targets

Pion electroproduction is described by six independent complex amplitudes $h_\pm^N$, $h_\pm^F$, h_0^N, h_0^F and therefore involves eleven independent quantities that must be measured for a completely model-independent determination of the process. The amplitudes $h_+^{N,F}$ ($h_-^{N,F}$) correspond to incident virtual photons polarized perpendicular (parallel) to the scattering plane; F and N refer to baryon flip and nonflip, respectively /205/; $h_0^{N,F}$ are longitudinal helicity amplitudes /206/.

As long as unpolarized leptons and unpolarized targets are used, only the four terms appearing in (1.13) can be determined, which are expressed by the following relations in terms of the amplitudes defined above:

$$\sigma_u = \frac{|\underline{q}^*|}{k_L} \frac{W}{m_N} \frac{1}{2} \left(|h_+^N|^2 + |h_+^F|^2 + |h_-^N|^2 + |h_-^F|^2 \right),$$

$$\sigma_T = \frac{|\underline{q}^*|}{k_L} \frac{W}{m_N} \frac{1}{2} \left(|h_-^N|^2 + |h_-^F|^2 - |h_+^N|^2 - |h_+^F|^2 \right),$$

$$\sigma_L = \frac{|\underline{q}^*|}{k_L} \frac{W}{m_N} \frac{1}{2} \left(|h_o^N|^2 + |h_o^F|^2 \right), \tag{6.25}$$

$$\sigma_I = \frac{|\underline{q}^*|}{k_L} \frac{W}{m_N} 2 \,\mathrm{Re} \left(h_o^N h_-^{N*} + h_o^F h_-^{F*} \right).$$

Therefore, a complete model-independent determination of electroproduction is only possible by using polarized beams and targets.

The problem has been discussed by a number of authors, a long list of which can be found in the paper by ACTOR /207/, who gives the general expressions for the cross sections for the following types of experiments:

$$l + a \to l' + X,$$

$$l + a \to l' + c + X, \tag{6.26}$$

$$l + a \to l' + c + d,$$

where the target particle a can have arbitrary spin and polarization, X denotes any hadronic system, and c and d hadrons whose momenta are measured.

Very useful for the experimenters, although less general, are the articles of DOMBEY /208/ on elastic and inelastic scattering of polarized leptons, and of BARTL and MAJEROTTO /206/ who consider electroproduction of a single pion with an accompanying recoiling nucleon in the final state.

Following the latter authors, we recall that, if the incident lepton is polarized (longitudinally), a fifth term should be added to (1.13)

$$\sigma_l = -\xi \frac{|\underline{q}^*|}{k_L} \frac{W}{m_N} \sqrt{2\varepsilon(1-\varepsilon)} \sin\phi^* \,\mathrm{Im} \left\{ h_o^N h_-^{N*} + h_o^F h_-^{F*} \right\}, \tag{6.27}$$

which is proportional to the imaginary part of the same function, whose real part appears in σ_I. In this expression

$$\xi = \frac{\underline{n}\cdot\underline{l}_1}{|\underline{l}_1|}$$

measures the degree of longitudinal polarization of the incident lepton. Its polarization is characterized by the 4-vector s^μ, which fulfills the relations $s^2 = -1$, $s\cdot l_1 = 0$ and reduces to $s^\mu \equiv (0, \underline{n})$ in the rest system of the lepton.

If only the target is polarized, the following term should be added to (1.13):

$$\sigma_t = \frac{|\underline{q}^*|}{k_L}\frac{W}{m_N}\Big\{ P_x \Big[-\sqrt{2\varepsilon(1+\varepsilon)}\,\sin\phi^*\,\mathrm{Im}\,X_1 - \varepsilon\sin 2\phi^*\,\mathrm{Im}\,X_2\Big] -$$
$$- P_y\Big[\mathrm{Im}\,Y_1 + \varepsilon\cos 2\phi^*\,\mathrm{Im}\,Y_2 + 2\varepsilon\,\mathrm{Im}\,Y_3 + \sqrt{2\varepsilon(1+\varepsilon)}\cdot\cos\phi^*\,\mathrm{Im}\,Y_4\Big] + \qquad (6.28)$$
$$+ P_z\Big[\varepsilon\sin 2\phi^*\,Z_2 + \sqrt{2\varepsilon(1+\varepsilon)}\,\sin\phi^*\,\mathrm{Im}\,Z_1\Big]\Big\}$$

where $\underline{P}$ is the polarization of the target and

$$X_1 = h_0^F h_+^{N*} + h_0^N h_+^{F*}, \qquad X_2 = h_-^F h_+^{N*} + h_-^N h_+^{F*},$$
$$Y_1 = h_+^N h_+^{F*} + h_-^N h_-^{F*}, \qquad Y_2 = h_-^N h_-^{F*} - h_+^N h_+^{F*},$$
$$Y_3 = h_0^N h_0^{F*}, \qquad Y_4 = h_0^N h_-^{F*} - h_0^F h_-^{N*}, \qquad (6.29)$$
$$Z_1 = h_0^N h_+^{N*} - h_0^F h_+^{F*}, \qquad Z_2 = h_-^N h_+^{N*} - h_-^F h_+^{F*}$$

CHRIST and LEE /209/ have shown that σ_t must vanish if time reversal invariance is valid. Thus a nonzero asymmetry of the distribution of the scattered electrons with respect to the plane orthogonal to the nucleon polarization would be a clear test of T-invariance violation. Such a conclusion involves only the validity of the one-photon exchange approximation.

Two experiments /210, 211/ on electron inelastic scattering from polarized protons in the region of resonance excitation Δ (1232), N^* (1512), N^* (1688) and $-k^2$ between 0.2 and 1 $(\mathrm{GeV}/c)^2$ did not reveal any sizeable violation of the time-reversal invariance.

Finally, if both electron and target are polarized, one must add to the cross section (1.13) a third term

$$\sigma_{lt} = -\xi\,\frac{|\underline{q}^*|}{k_L}\frac{W}{m_N}\Big\{ -P_x\big[\sqrt{2\varepsilon(1-\varepsilon)}\,\cos\phi^*\,\mathrm{Re}\,X_1 + \sqrt{1-\varepsilon^2}\,\mathrm{Re}\,X_2\big] +$$
$$+ P_y\sqrt{2\varepsilon(1-\varepsilon)}\,\sin\phi^*\,\mathrm{Re}\,Y_4 + P_z\big[\sqrt{1-\varepsilon^2}\,\mathrm{Re}\,Z_2 + \sqrt{2\varepsilon(1-\varepsilon)}\,\cos\phi^*\,\mathrm{Re}\,Z_1\big]\Big\}. \qquad (6.30)$$

It is interesting to note that, using just polarized targets, only imaginary parts of products of amplitudes are measured, whereas additional use of polarized leptons gives us information about the corresponding real parts. Note also that while the ϕ^* dependence of (1.13) is given by $\cos\phi^*$ and $\cos 2\phi^*$, in the new expressions σ_l, σ_t and σ_{lt} there are also terms with $\sin\phi^*$ and $\sin 2\phi^*$ factors.

It is then possible to define a number of asymmetries (varying between +1 and -1), the determination of which requires the measurement of cross sections for different values of ϕ^*.

As an illustration of these formulae, the asymmetries can be estimated for the pure Born term model (N and π exchange), which, with some modifications /212/, is in reasonable agreement with the existing data for $\sigma_U + \varepsilon\sigma_L$, σ_T, and σ_I. In its pure form it gives only real amplitudes and, therefore, nonzero asymmetries only when both the electron and target proton are polarized. The reader is referred to the original paper /206/ for numerical results.

Appendix A: Kinematical Relations

A.1 Definitions and General Relations[32]

The conservation of energy and momentum for reactions (1.1) is expressed by the relation among four vectors

$$l_1 + p_1 = l_2 + p_2 + q,$$

where l_1, l_2 refer to the initial and final lepton (of mass μ), p_1, p_2 to the initial and final nucleon (of mass m_N), and q to the produced meson (of mass m_π) (Fig. 1.1). Introducing the four-momentum

$$k = l_1 - l_2 \tag{A.1}$$

of the virtual photon exchanged between the leptonic and hadronic vertices, the conservation of energy and momentum becomes

$$k + p_1 = p_X = p_2 + q, \tag{A.2}$$

where p_X is the four-momentum of the recoiling hadron πN system which can be considered as an intermediate state or particle X that decays into a pion and a nucleon.

The square of the four-momentum (A.1) is given by

$$k^2 = k_0^2 - \underline{k}^2 = 2\,\mu^2 - 2\,l_{01}\,l_{02}\,(1-\beta_1\,\beta_2\,\cos\theta_1), \tag{A.3}$$

which, for l_{01}, $l_{02} >> \mu$, reduces to the commonly used expression

$$k^2 = -4\,l_{01}\,l_{02}\,\sin^2(\theta_1/2). \tag{A.4}$$

32 The material covered in Appendices A, B, C is discussed in several review articles. See for instance /23, 61, 82, 208/.

As we see from this relation, in electroproduction the four-momentum transfer is spacelike, $k^2 < 0$, and its square represents the square of the (imaginary) mass of the virtual photon.

It can be computed in terms of quantities measured in the l.f. by means of (A.3) or, usually, (A.4).

Other important invariants are the usual Mandelstam variables

$$\begin{aligned} s &= W^2 = (p_1+k)^2 = m_N^2 + k^2 + 2\, p_1\cdot k = m_N^2 + m_\pi^2 + 2\, p_2\cdot q, \\ t &= (k-q)^2 = k^2 + m_\pi^2 - 2\, q\cdot k = 2\, m_N^2 - 2\, p_1\cdot p_2, \\ \bar{s} &= (p_2-k)^2 = k^2 + m_N^2 - 2\, p_2\cdot k = m_\pi^2 + m_N^2 - 2\, p_1\cdot q, \end{aligned} \tag{A.5}$$

fulfilling the well-known relation

$$s + \bar{s} + t = k^2 + 2\, m_N^2 + m_\pi^2.$$

Very often the variable t is written in the form

$$t = t_{min} - 4\, |\underline{k}|\,|\underline{q}|\, \sin^2 (\theta_{kq}/2), \tag{a}$$

where (A.6)

$$t_{min} = (k_o-q_o)^2 - (|\underline{k}| - |\underline{q}|)^2. \tag{b}$$

Other variables frequently used are

$$\begin{aligned} &P = \tfrac{1}{2}\,(p_1 + p_2), \quad \Delta = p_2 - p_1, \qquad P\cdot\Delta = 0, \\ &\nu = q\cdot P = k\cdot P, \\ &\nu_B = -\tfrac{1}{2}\, q\cdot k. \end{aligned} \tag{A.7}$$

These are related to s and $\bar{s}$ as follows:

$$\begin{aligned} s &= M^2 + 2\, (\nu-\nu_B), \\ \bar{s} &= M^2 - 2\, (\nu+\nu_B). \end{aligned} \tag{A.8}$$

A.2 The Same Quantities in Specific Reference Frames

The most frequently used reference frames are: the frame of the c.m. (c.m.f.) of the final πN system, the laboratory frame (l.f.), and three Breit frames defined below.

The c.m.f. is defined by the relation

$$\underline{p}_1^* = -\underline{k}^* \quad \text{or} \quad \underline{p}_2^* = -\underline{q}^*, \tag{A.9}$$

and one has the useful formulae

$$\begin{aligned}
k_o^* &= \frac{W^2+k^2-m_N^2}{2\,W}, \qquad q_o^* = \frac{W^2+m_\pi^2-m_N^2}{2\,W},\\
\underline{k}^{*2} &= \left(\frac{W^2+k^2-m_N^2}{2m_N}\right)^2 - k^2 = \left(\frac{W^2+m_N^2-k^2}{2\,W}\right)^2 - m_N^2,\\
\underline{q}^{*2} &= \left(\frac{W^2+m_\pi^2-m_N^2}{2\,W}\right)^2 - m_\pi^2 = \left(\frac{W^2+m_N^2-m_\pi^2}{2\,W}\right)^2 - m_N^2,\\
p_{01}^* &= W - k_o^* = \frac{W^2+m_N^2-k^2}{2\,W}\,.\\
p_{02}^* &= W - q_o^* = \frac{W^2+m_N^2-m_\pi^2}{2\,W}
\end{aligned} \tag{A.10}$$

At threshold $W = m_N+m_\pi$, i.e., $\underline{q}^* = \underline{0}$, which gives for the invariant variables

$$\begin{aligned}
(t)_{th.} &= 2\,m_N^2 - 2\,m_N\,(p_{01}^*)_{th.} = (k^2-m_\pi^2)\,(1+m_\pi/m_N)^{-1},\\
(\nu)_{th.} &= \frac{m_\pi}{2}\,(m_N+p_{01}^*)_{th.} = m_\pi\,m_N\,(1-t_{th}/4m_N^2).
\end{aligned} \tag{A.11}$$

The l.f. is defined by the relations

$$\underline{p}_1 = \underline{0}, \qquad p_{01} = m_N, \tag{A.12}$$

so that one obtains the following useful relations:

$$\begin{aligned}
s &= W^2 = k^2 + m_N^2 + 2\,m_N\,k_o,\\
t &= \Delta^2 = 2\,m_N^2 - 2\,m_N\,p_{02} = -2\,m_N\,T_2,
\end{aligned} \tag{A.13}$$

where $T_2 = p_{02} - m_N$ is the kinetic energy of the recoiling nucleon.

Instead of s, sometimes $W = \sqrt{s}$ or the photon equivalent energy

$$k_L = \frac{s-m_N^2}{2m_N} \tag{A.14}$$

is used: k_L is the energy that a real photon ($k^2 = 0$) must have in the l.f. in order to produce the final πN system with the same invariant mass, W.

According to (A.2), the momentum $\underline{p}_X$ of the recoiling intermediate hadron is equal to $\underline{k}$. The angle θ_X between $\underline{p}_X$ and $\underline{l}_1$ is given by

$$\sin\theta_X = \frac{|\underline{l}_2|}{|\underline{k}|}\sin\theta_1. \tag{A.15}$$

The Lorentz transformation from the l.f. to the c.m.f. is

$$\underline{\beta} = \frac{\underline{k}}{k_o+m_N}, \qquad \gamma = \frac{k_o+m_N}{W},$$
$$k_o^* = \frac{m_N k_o + k^2}{W}, \quad \underline{k}^* = \frac{m_N}{W}\underline{k}, \tag{A.16}$$

and $\underline{k}$ and $\underline{k}^*$ are parallel.

The intermediate state X decays into a pion and a nucleon. Just above threshold, both these particles are emitted in the forward direction. By increasing W, the pion velocity in the c.m.f. fairly soon reaches the value of the velocity of the c.m. and therefore is emitted over the whole 4π solid angle in the l.f. The nucleon is emitted only inside a cone, the axis of which is defined by (A.15). Its semi-aperture θ_{max} is given by

$$\cos\alpha_2 < \cos\theta_{max} = \frac{1}{\beta}\left(1 - \gamma p_o^{*2}/m_N^2\right), \tag{A.17}$$

where β and γ are given by (A.16) and p_{02}^* by (A.10). In each direction within this cone tne heavy hadron momentum has two values

$$|\underline{p}_2| = \frac{p_{02}^*\,\beta\cos\alpha_2 \pm \left[p_{02}^{*2} - m_N^2\gamma^2\,(1-\beta^2\cos^2\alpha_2)\right]^{1/2}}{\gamma(1-\beta^2\cos^2\alpha_2)}. \tag{A.18}$$

Finally three "Breit frames" are used advantageously in certain problems.

The *nucleon Breit frame* (n.B.f.) is defined by the relation

$$\underline{P}^N = \underline{0}, \quad \underline{p}_2^N = -\underline{p}_1^N = \underline{\Delta}^N/2, \tag{A.19}$$

so that

$$\Delta_0^N = p_{02}^N - p_{01}^N = 0, \quad |\underline{\Delta}^N| = \sqrt{-t},$$
$$p_{01}^N = p_{02}^N = \sqrt{m_N^2 - t/4}\ . \tag{A.20}$$

The *hadron Breit frame* (h.B.f.) is defined by the relations

$$p_{0X}^h = p_{01}^h \quad \text{and} \quad \underline{p}_X^h \,||\, -\underline{p}_1^h.$$
$$|\underline{p}_X^h| = \sqrt{\underline{p}_1^{h2} - (W^2 - m_N^2)} < |\underline{p}_1^h|,$$
$$k_0^h = p_{0X}^h - p_{01}^h = 0, \quad \underline{k}^h \,||\, \underline{p}_1^h, \tag{A.21}$$
$$k^2 = -\underline{k}^{h^2}.$$

The behaviour of the electron is the same as if it would be elastically rebound on a rigid wall (Fig. A.1) with an angle of deflection θ_e^h given by

$$\sin\frac{\theta_e^h}{2} = \frac{\sqrt{-k^2}}{2l_1^h}\ . \tag{A.22}$$

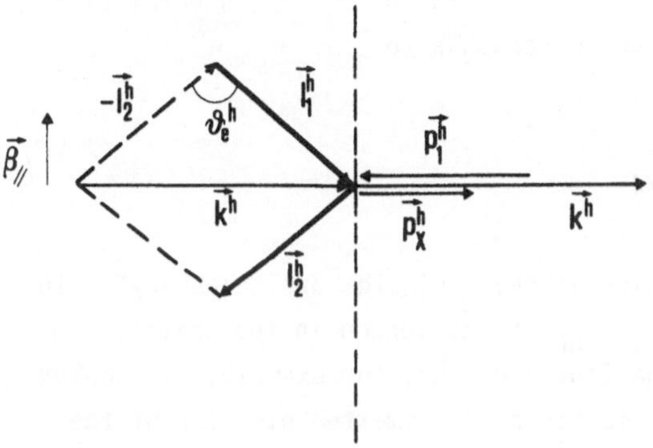

Fig. A.1. Hadron Breit frame

The *lepton Breit frame* (l.B.f.) is defined by the relation

$$\underline{l}_2 = -\underline{l}_1, \tag{A.23}$$

so that

$$l_{O2}^l = l_{O1}^l, \quad k_0^l = 0, \quad \underline{k}^l = 2\,\underline{l}_1^l = \underline{p}_X^l - \underline{p}_1^l, \quad p_{0X}^l = p_{01}^l. \tag{A.24}$$

In this frame the hadronic system does not receive energy from the lepton current and behaves as if it would elastically rebound (Fig. A.2).

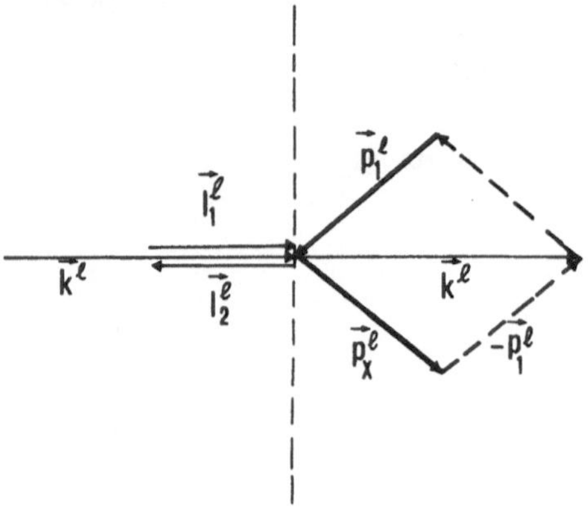

Fig. A.2. Lepton Breit frame

The Lorentz transformation from the c.m.f. to the h.B.f. reduces k_0 to $k_0^h = 0$ and therefore

$$\beta = \frac{k_0^*}{|\underline{k}^*|} , \qquad \gamma = \frac{|\underline{k}^*|}{\sqrt{-k^2}} . \tag{A.25}$$

Finally the Lorentz transformation from the h.B.f. to the l.B.f. is parallel to the brick wall, i.e., transversal to $\underline{k}^h$, and corresponds to

$$\beta = \frac{|\underline{l}_1^h|\cos(\theta_1^h/2)}{l_{01}^h} , \qquad \gamma = \frac{l_{01}^h}{\sqrt{l_{01}^{h2}+k^2/4}} . \tag{A.26}$$

To specify one event completely, five kinematical variables are necessary[33]. In the l.f. one normally uses the energies l_{01}, l_{02} of the lepton in the initial and final state, the angle of scattering of the lepton θ_l and, for example, two angles (θ_π and ϕ_π) which specify the direction of motion of the emitted pion (or of the recoiling nucleon).

[33] The five particles appearing in electroproduction of a single boson (2 in the initial and 3 in the final state) correspond to 15 degrees of freedom, 3 of which refer to the rotations and 3 to the translations of the system as a whole. Of the remaining 9 degrees of freedom only 5 correspond to independent variables because of the 4 relations (A.2) expressing the conservation of energy and momentum.

The most significant variables from the physical point of view are: the invariants k^2 and s (or W or k_L), the polarization parameter ε [(A.29)] and the angles θ_π^*, ϕ_π^* that define the direction of motion of the pion in the c.m.f. Sometimes instead of θ_π^* one uses t.

The first three variables k^2, s (W or k_L), and ε completely describe the properties of the virtual photon. They are determined by the electron channel since they depend only on l_{01}, l_{02} and θ_1. The knowledge of the two remaining variables requires the observation of one of the two hadrons in coincidence with the inelastically scattered lepton.

A.3 The Polarization of the Virtual Photon

The polarization of the virtual photon (of nonzero mass: k^2) depends on the reference frame. To clarify this point, let us consider the lepton Breit frame, where the final lepton recoils with a momentum $\underline{l}_2^1$ opposite to the initial one.

In this frame unpolarized electrons give rise to a 50 % mixture of photons with helicity ± 1 and no population in the helicity 0 state. When, by a Lorentz transformation, we pass to the c.m.f., the photon acquires the transverse polarization

$$\varepsilon = \frac{|A_x^*|^2 - |A_y^*|^2}{|A_x^*|^2 + |A_y^*|^2} = \frac{\rho_{xx}-\rho_{yy}}{\rho_{xx}+\rho_{yy}}, \tag{A.27}$$

and the longitudinal polarization

$$\varepsilon_L = \frac{|A_z^*|^2}{|A_x^*|^2 + |A_y^*|^2} = \frac{\rho_{zz}}{\rho_{xx}+\rho_{yy}} = -\frac{k^2}{k_0^{*2}}\,\varepsilon, \tag{A.28}$$

where A_i^* are the components of the vector potential of the virtual photon and ρ_{ij} is the photon polarization density matrix defined by (B.12). The c.m. reference frame is shown in Fig. A.3; the z-axis is taken in the direction of the three-momentum of the virtual photon

$$\hat{z} = \hat{k}^*,$$

the y-axis perpendicular to the scattering plane,

$$\hat{y} = \hat{l}_1^* \times \hat{l}_2^* \, (\sin\theta_1^*)^{-1}$$

and the x-axis in the scattering plane

$$\hat{x} = \hat{y} \times \hat{z}.$$

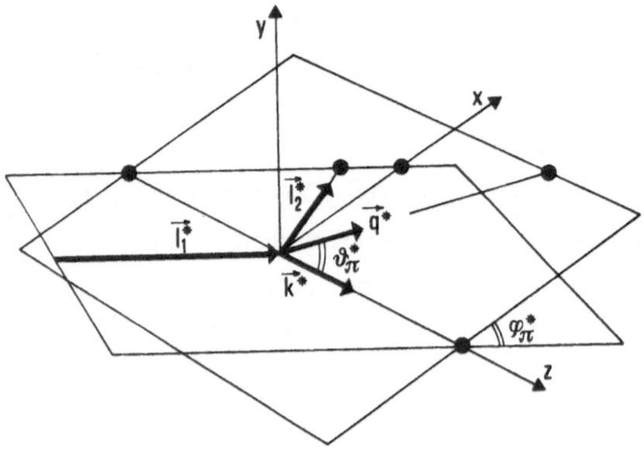

Fig. A.3. Electroproduction coordinates in the c.m. frame of final hadrons

The *polarization parameter* ε can easily be expressed in terms of quantities measured in the l.f. (Section B.2)

$$\varepsilon = \left[1 + 2\,\frac{\underline{k}^2}{-k^2}\;tg^2\;(\theta_1/2)\right]^{-1}. \tag{A.29}$$

It can easily be shown that, for fixed values of k^2 and W, the parameter ε is a decreasing function of the scattering angle θ_1 (Fig. A.4).

As shown by (A.28) the longitudinal polarization parameter can be expressed in terms of ε, so that, for fixed values of k^2 and W, it also decreases by increasing θ_1.

From (A.16) it follows that the Lorentz transformations from the l.f. to the c.m.f. (or the h.B.f.) are parallel to the direction of $\underline{k}^*$ which has been taken as direction of the z-axis (Fig. A.3). Therefore, although not a relativistic invariant, the polarization parameter ε is invariant with respect to the transformations from one to the other of these frames, since, according to its definition (A.27), it involves only the transversal components of the four-vector $\underline{A}^*$.

A.4 Kinematics of the Inverse Reaction

In the case of reaction (6.1) considered in Section 6, instead of (A.1), (A.2), one has (Fig. A.5)

$$\begin{aligned} k &= l_1 + l_2 \\ p_1 + q &= p_2 + k, \end{aligned} \tag{A.30}$$

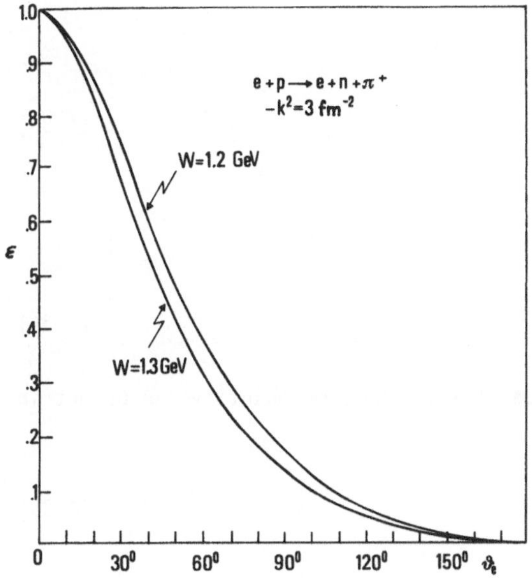

Fig. A.4. Polarization parameter versus angle of scattering of the electron at $-k^2 = 3\text{fm}^{-2}$ and W = 1.2 and 1.3 GeV

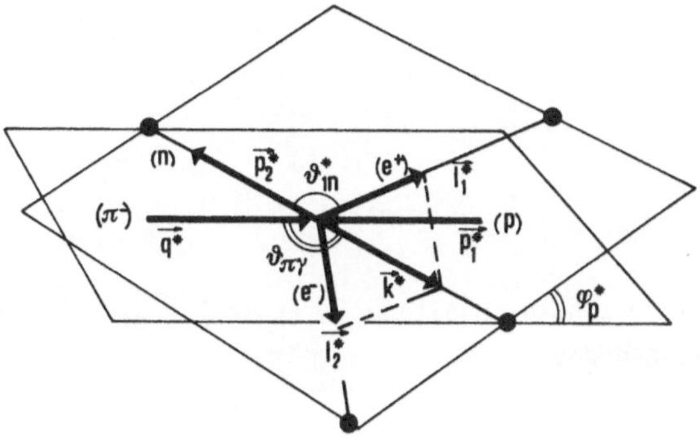

Fig. A.5. Inverse electroproduction: frame of the c.m. of initial hadrons

from the first of which it follows

$$k^2 = 2\,\mu^2 + 2\,(l_{01}\,l_{02} - \underline{l}_1\cdot\underline{l}_2) = 2\,\mu^2 + 2l_{01}\,l_{02}\,(1-\beta_1\beta_2\cos\theta_{12}) \approx$$
$$\approx 4\,l_{01}\,l_{02}\,\sin^2(\theta_{12}/2), \tag{A.31}$$

where θ_{12} is the angle between the two leptons (of opposite charge) present in the final state. As it appears from (A.31) k_μ, in this case, is always timelike ($k^2>0$).

A kinematical configuration of great interest is provided by the so-called *quasi-threshold condition*, which corresponds to imposing, in the c.m.f., that the three-momentum of the virtual photon be zero

$$\underline{k}^* = \underline{0}. \tag{A.32}$$

Under this condition (Fig. A.6)

$$\underline{l}_2^* = -\underline{l}_1^*, \qquad l_{01}^* = l_{02}^*,$$
$$\underline{p}_2^* = \underline{0}, \qquad k_0^* = 2\, l_{01}^*, \tag{A.33}$$

and the four-momentum transfer

$$k^2 = 4\, l_{01}^{*2} \tag{A.34}$$

reaches its maximum value for a fixed value of l_{01}^*. The same quantity can be put in two forms

$$\sqrt{k^2} = \sqrt{k_0^{*2}} = q_0^* + p_{01}^* - p_{02}^* = \sqrt{s} - m_N. \tag{A.35}$$

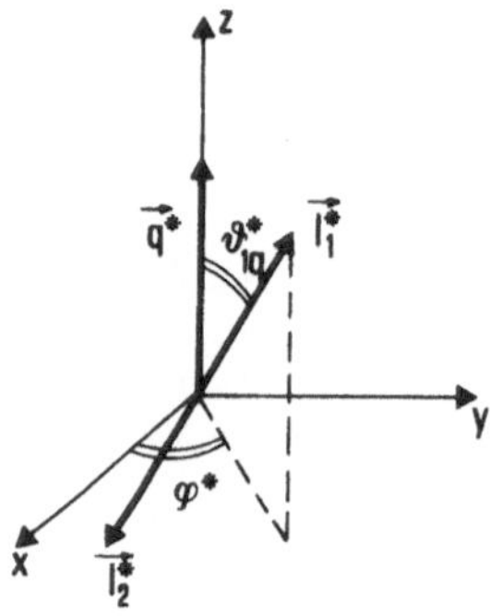

Fig. A.6. Quasi-threshold condition in the frame of the c.m. of initial hadrons

Furthermore, if the specification of the final state is made by assigning, besides the angle θ_{12}^*, the angles θ_{1n}^* between the positron and the recoiling nucleon and ϕ_p^* between the reaction plane (defined by the vectors $\underline{p}_1^*$ and $\underline{p}_2^*$) and the photon decay plane (defined the by vectors $\underline{l}_1^*$ and $\underline{l}_2^*$), at quasithreshold one has

$$\theta_{1n}^* = 0, \quad \phi_p^* = 0. \tag{A.36}$$

The quasithreshold condition should be introduced with some caution, however, because where (A.36) is fulfilled, the angle θ_{kq}^* becomes indeterminate. A recipe for avoiding this indetermination consists of imposing first $\theta_{kq}^* = 0$ and then $\underline{k}^* = 0$.

Appenpix B: The Leptonic Part of the Differential Cross Section

B.1 The Electroproduction Differential Cross Section

The differential cross section of reaction (1.1) can be written in the form[34]

$$d^9\sigma = \frac{1}{(2\pi)^5\, 4\sqrt{(p_1 l_1)^2 - \mu^2 m_N^2}}\, \Sigma' \, |M_{fi}|^2 \, \frac{d\underline{l}_2 d\underline{p}_2\, d\underline{q}}{2l_{02} 2p_{02} 2q_0} \quad \delta(p_1 + k - p_2 - q), \tag{B.1}$$

where Σ' sums over final spin states and averages over initial spin states.

In the o.p.e.a. the transition matrix element is given by

$$M_{fi} = \frac{e^2}{k^2}\, \bar{u}(l_2)\, \gamma_\mu\, u(l_1) < p_2\, q|V_\mu|p_1> \equiv \frac{e^2}{k^2}\, l_\mu\, M^\mu, \tag{B.2}$$

where V_μ is the hadron current and M_μ is the physical quantity which we are interested to obtain from experiments. The cross section (B.1) involves the quantity

$$\Sigma' \, |M_{fi}|^2 = \frac{2e^4}{k^4}\, L_{\mu\nu}\, M^{\mu\nu}, \tag{B.3}$$

where, for convenience, we define

$$L_{\mu\nu} = \frac{1}{4}\, \Sigma\, l_\mu\, l_\nu^*, \quad M_{\mu\nu} = \frac{1}{2}\, \Sigma\, M_\mu\, M_\nu^*. \tag{B.4}$$

By standard trace techniques, we obtain for unpolarized leptons

$$L_{\mu\nu} = \frac{1}{2}\, k^2\, g_{\mu\nu} + l_{1\mu}\, l_{2\nu} + l_{1\nu}\, l_{2\mu}. \tag{B.5}$$

From the Lorentz condition $k^\mu\, l_\mu = 0$, i.e., $l_0 = \underline{k}\cdot\underline{l}/k_0$, and $k^\mu L_{\mu\nu} = L_{\mu\nu} k^\nu = 0$.

---

[34] We adopt the covariant normalization

$<p_2|p_1> = 2E(2\pi)^3\, \delta(\underline{p}_2 - \underline{p}_1)$.

In particular for spinors

$\bar{u}u = 2m_N$, $u^+u = 2\, E$.

Similarly, the conservation of hadronic current gives $k^{\mu}M_{\mu} = 0$, i.e., $M_0 = \underline{k}\cdot\underline{M}/k_0$, and

$$k^{\mu}M_{\mu\nu} = M_{\mu\nu}k^{\nu} = 0.$$

These relations imply that it is sufficient to consider only space components of $L_{\mu\nu}$ and $M_{\mu\nu}$. With the convention of taking the z-axis in the direction of the vector $\underline{k}$, we obtain

$$\begin{aligned} l^{\mu}M_{\mu} &= l_o\,M_o - \underline{l}\cdot\underline{M} = -l_x\,M_x - l_y\,M_y - (1-k^2/k_0^2)\,l_z\,M_z = \\ &= -(l_x\,M_x + l_y\,M_y + \frac{k^2}{k_0^2}\,l_z\,M_z). \end{aligned} \tag{B.6}$$

We can include the current conservation factor k^2/k_0^2 in the lepton contribution to (B.2) and ignore it in the hadronic part.

The same factor that compensates the elimination of the time components of the currents can be used for transforming the scalar (or time) part of the cross section into a longitudinal part. (A similar remark holds also for the terms of the development in multipoles of the electroproduction maplitudes.)

B.2 The Density Matrix and the Polarization Parameters

According to the definitions (A.27) and (A.28), we can write

$$\varepsilon = \frac{\rho_{xx}-\rho_{yy}}{\rho_{xx}+\rho_{yy}}, \quad \varepsilon_L = \frac{\rho_{zz}}{\rho_{xx}+\rho_{yy}}, \tag{B.7}$$

where the 3 x 3 matrix

$$\rho_{ij} = -\frac{1}{k^2}\left(-\frac{k^2}{k_0^2}\right)^{\delta_{iz}+\delta_{jz}} L_{ij} \qquad (i,j = x,\,y,\,z) \tag{B.8}$$

is the virtual photon polarization density matrix. The factor on the right-hand side compensates for the elimination of the time components (in a given reference frame).

To compute ρ_{xx} and ρ_{yy}, we consider the triangle of sides $\underline{k}$, $\underline{l}_1$, $\underline{l}_2$ and take the z-axis in the direction of $\underline{k}$ and the y-axis perpendicular to the plane of the triangle.

Then we obtain from (B.5)

$$L_{xx} = -\frac{1}{2}\,k^2 + 2l_{1x}\,l_{2x}, \quad L_{yy} = -\frac{1}{2}\,k^2 + 2l_{1y}\,l_{2y}$$

where

$$l_{1y} = l_{2y} = 0 \quad l_{1x} = l_{2x} = \frac{l_1 l_2}{|\underline{k}|} \sin \theta_1 .$$

From (A.4) one has further

$$l_1 \cdot l_2 \approx - \frac{k^2}{4\sin^2\theta_1} ,$$

so that

$$\rho_{xx} = \frac{1}{2} - \frac{k^2}{2\underline{k}^2} \operatorname{cotg}^2 (\theta_1/2), \qquad \rho_{yy} = \frac{1}{2}. \tag{B.9}$$

These relations lead us finally to the result

$$\varepsilon = (1 - 2 \frac{\underline{k}^2}{k^2} \operatorname{tg}^2 \theta_1)^{-1}. \tag{B.10}$$

The quantity ε is thus purely determined by the lepton kinematics and it can be computed in any reference frame using the above general expression. Similarly, using the

$$l_{1z} = \frac{l_1}{|\underline{k}|} (l_1 - l_2 \cos \theta_1), \quad l_{2z} = \frac{l_2}{|\underline{k}|} (l_1 \cos \theta_1 - l_2),$$

one obtains for the z components

$$\rho_{zz} = \left(\frac{k^2}{k_o^2}\right)^2 \frac{k_o^2}{2\underline{k}^2} \operatorname{cotg}^2 \theta_1/2 = - \frac{k^2}{k_o^2} \frac{\varepsilon}{1-\varepsilon} , \ \varepsilon_L = - \frac{k^2}{k_o^2} \varepsilon,$$
$$\rho_{xz} = - \frac{1}{1-\varepsilon}\sqrt{\frac{\varepsilon_L(1+\varepsilon)}{2}} . \tag{B.11}$$

Thus we finally get the explicit form of the photon polarization density matrix

$$\rho = \frac{1}{1-\varepsilon} \begin{vmatrix} \frac{1+\varepsilon}{2} & 0 & -\sqrt{\frac{\varepsilon_L(1+\varepsilon)}{2}} \\ 0 & \frac{1-\varepsilon}{2} & 0 \\ -\sqrt{\frac{\varepsilon_L(1+\varepsilon)}{2}} & 0 & \varepsilon_L \end{vmatrix} . \tag{B.12}$$

In order to be convinced that ε, as determined by (B.10), represents a measure of the transverse linear polarization of the virtual photon, one can compare (B.12) with the density matrix of real, transverse photons. If the real photon propagates along the z-axis, ρ_{ij} has only x, y components, and for a partially linearly polarized beam of relative strength $1+\varepsilon$ to $1-\varepsilon$ one has

$$\rho = \frac{1}{1-\varepsilon} \begin{vmatrix} \frac{1}{2}(1+\varepsilon) & 0 & 0 \\ 0 & \frac{1}{2}(1-\varepsilon) & 0 \\ 0 & 0 & 0 \end{vmatrix} , \qquad \text{(B.12')}$$

which is precisely (B.12) as $k^2 = 0$.

The case $\varepsilon = 0$ corresponds to unpolarized photons and after (B.10) this occurs for $\theta_1 = \pi$, i.e., backward scattering. More generally, this configuration defines the lepton Breit frame. When the lepton is scattered in the backward direction, since only spin-flip transitions can produce spin 1 photons, the emitted virtual photon has to carry off helicity ± 1 and is therefore circularly polarized. If, in particular, the leptons are unpolarized, there is no net photon polarization. The longitudinal photon polarization then arises as a consequence of the Lorentz transformation from the l.B.f. to l.f. or c.m.f.

B.3 The Matrix Elements in h.B.f. and c.m.f.

To evaluate the cross section, although it is possible to read it from the density matrix, it is simpler first to evaluate the quantity $L^{\mu\nu} M_{\mu\nu}$ in the h.B.f. and then to Lorentz transform it to the c.m.f.

Consider $L_{\mu\nu}$: the vectors $l_{1,2}$ have components

$$\left(l^h_{1,2}\right)_\mu \equiv \left(l^h,\ l^h \cos(\theta^h_1/2) \cos\phi,\ \ l^h \cos(\theta^h_1/2) \sin\phi, \pm l^h \sin(\theta^h_1/2)\right) ,$$

where

$$l^h = \frac{1}{2} \sqrt{-k^2} \operatorname{cosec}(\theta^h_1/2),$$

and ϕ is the angle between the lepton scattering plane $(\hat{l}^h_1, \hat{l}^h_2)$ and the final hadron plane.

Using the above expressions we obtain

$$
\begin{aligned}
L_{oo} &= -\frac{1}{2}k^2 \operatorname{cotg}^2(\theta_1^h/2),\\
L_{xx} &= -\frac{1}{2}k^2\left[1+\operatorname{cotg}^2(\theta_1^h/2)\cos^2\phi\right],\\
L_{yy} &= -\frac{1}{2}k^2\left[1+\operatorname{cotg}^2(\theta_i^h/2)\sin^2\phi\right],\\
L_{ox} = L_{xo} &= -\frac{1}{2}k^2 \operatorname{cosec}(\theta_1^h/2)\operatorname{cotg}(\theta_1^h/2)\cos\phi,\\
L_{\ z} = L_{z\ } &= 0.
\end{aligned}
\tag{B.13}
$$

No other components of $L_{\mu\nu}$ are required, since the only contributions to $L^{\mu\nu}M_{\mu\nu}$ from the symmetric part of $M_{\mu\nu}$ contain terms in $g_{\mu\nu}$ and tensors formed by p_1, k, q. None of these vectors has a y-component so that

$$M_{oy} + M_{yo} = M_{xy} + M_{yx} = 0.$$

Hence

$$L_{\mu\nu}M^{\mu\nu} = L_{xx}M_{xx}^h + L_{yy}M_{yy}^h + L_{oo}M_{oo}^h + L_{ox}(M_{ox}+M_{xo}). \tag{B.14}$$

The matrix element $M_{\mu\nu}$ must be Lorentz transformed from the h.B.f. to the c.m.f., which is the most convenient for its evaluation. As already discussed in Appendix A, such a Lorentz transformation, leading from $k_\mu^h \equiv (0, 0, 0, k^h)$ to $k_\mu^* \equiv (k_0^*, 0, 0, k^*)$, acts in the z-direction and leaves unchanged the transversal components. Since

$$J_0^h = \gamma(J_0^* - \beta J_3^*) = \left(\frac{k^{*2}}{k_0^*\sqrt{-k^2}} - \frac{k_0^*}{\sqrt{-k^2}}\right)J_3^* = -\frac{\sqrt{-k^2}}{k_0^*}J_3^*,$$

we obtain

$$M_{ox}^h = \frac{\sqrt{-k^2}}{k_0^*}M_{zx}^*,\quad M_{oo}^h = -\frac{k^2}{k_0^{*2}}M_{zz}^*$$

and (B.14) becomes

$$
\begin{aligned}
\frac{2}{-k^2}L^{\mu\nu}M_{\mu\nu} &= \frac{M_{xx}^*+M_{yy}^*}{2}\left[2+\operatorname{cotg}^2(\theta_1^h/2)\right] + \frac{-k^2}{k_0^{*2}}M_{zz}^*\operatorname{cotg}^2(\theta_1^h/2) +\\
&+ \frac{M_{xx}^*-M_{yy}^*}{2}\operatorname{cotg}^2(\theta_1^h/2)\cos 2\phi + (M_{zx}^*+M_{xz}^*)\frac{\sqrt{-k^2}}{k_o^*}\operatorname{cosec}(\theta_1^h/2)\operatorname{cotg}(\theta_1^h/s)\cos\phi.
\end{aligned}
\tag{B.15}
$$

On the other hand, evaluating ε, given by (B.10), in this particular frame ($k_0^h = 0$, $k^2 = -\underline{k}^{h^2}$),

$$\varepsilon = \left[1 + 2\ tg^2\ (\theta_1^h/2)\right]^{-1} : \tag{B.16}$$

Since ε remains unchanged in the Lorentz transformation from h.B.f. to c.m.f. or to l.f., this relation can be used to compute the angle θ_1^h from quantities measured in these frames.

From (B.16) we obtain

$$\cot g^2\ (\theta_1^h/2) = \frac{2\varepsilon}{1-\varepsilon}\ ,\quad \operatorname{cosec}^2\ (\theta_1^h/2) = \frac{1+\varepsilon}{1-\varepsilon}\ ,$$

which allows us to give $L^{\mu\nu}\ M_{\mu\nu}$ the final form

$$L_{\mu\nu}\ M^{\mu\nu} = -\frac{k^2}{1-\varepsilon}\left[\frac{M_{xx}^*+M_{yy}^*}{2} + \varepsilon_L\ M_{zz}^* + \frac{M_{xx}^*-M_{yy}^*}{2}\ \varepsilon\cos 2\phi + \right. \left. + \frac{M_{zx}^*+M_{xz}^*}{2}\sqrt{2\ \varepsilon_L\ (1+\varepsilon)}\ \cos\phi\right] \tag{B.17}$$

where the matrix elements M_{ij}^* are computed in the centre-of-mass and depend on W, k^2, t (or θ^*). The azimuthal dependence on ϕ is, on the contrary, completely fixed and reproduces the result anticipated in (1.13). (The reason for selecting the explicit factors $\sin^2\theta^*$, $\sin\theta^*$ in that formula is a matter of convenience, related to the use of the c.m. amplitudes discussed in Appendix C.)

Let us go finally back to the original definition (B.1) for $d^9\sigma$. A standard calculation leads to the expression (1.11) of the cross section

$$\frac{d^5\sigma}{dl_{02}d\Omega_1 d\Omega_\pi^*} = \frac{\alpha}{2\pi^2}\frac{l_{02}}{l_{01}}\left(-\frac{1}{k^2}\right)\frac{K_L}{1-\varepsilon}\frac{d\sigma_v}{d\Omega_\pi^*} \tag{B.18}$$

where K_L is the photon equivalent energy [see (A.14)] and we have defined

$$\frac{d\sigma_v}{d\Omega_\pi^*} = \frac{\alpha}{16\pi}\frac{|\underline{q}^*|}{K_L}\frac{1}{m_N W}\left\{\frac{1-\varepsilon}{-k^2} L_{\mu\nu}\ M^{\mu\nu}\right\} . \tag{B.19}$$

The reason for this particular definition is to exhibit a form similar to the photoproduction cross section for a partially linearly polarized photon beam, which in the C.M. frame reads

$$\frac{d\sigma_\gamma}{d\Omega_\pi^*} = \frac{\alpha}{16\pi}\frac{|\underline{q}^*|}{|\underline{k}|}\frac{1}{m_N W}\left\{\frac{M_{xx}^* + M_{yy}^*}{2} + \varepsilon\ \frac{M_{xx}^* - M_{yy}^*}{2}\right\}. \tag{B.20}$$

In (B.20), $|\underline{k}|$ is the laboratory momentum of the real photon and that case $|\underline{k}| = K_L$.

Appendix C: Multipole Expansion

In Section 3 the general electroproduction amplitude has been analysed in terms of the relativistic invariant functions M_i (i = 1, 2 ... 8) defined in (3.12) and (3.15). The functions M_i are connected by the two constraints (3.24) due to current conservation.

Such a decomposition is particularly useful, since most properties relevant for a dispersion analysis can be simply expressed in its framework. The functions M_i have indeed simple analytic, crossing and asymptotic properties.

On the other hand, the invariant amplitudes M_i are not particularly suited to represent the physical constraints due to unitarity which, as discussed in Section 3, is particularly important in the analysis of low-energy electroproduction. In the energy region where single-pion production is dominant, it leads to the celebrated Fermi-Watson theorem stating that for each multipole the electroproduction amplitude is a complex quantity whose phase is equal to the corresponding pion-nucleon phase shift.

As a consequence the unitarity constraints become particularly simple and evident if one uses a "multipole" representation of the electroproduction amplitude in which:

1) One works explicitly in the centre-of-mass ystem.
2) One uses a gauge in which the virtual photon has only space components.
3) The fundamental kinematical variables are energy, angular momentum, and parity.

Our first step will be to exhibit the six independent c.m. amplitudes, functions of the total energy W and of the scattering angle θ

$$W = p_{01} + k_0 = p_{02} + q_0, \quad z = \cos\theta = \hat{k}\cdot\hat{q}, \tag{C.1}$$

where

$$\hat{k} = \underline{k}/|\underline{k}|, \quad \hat{q} = \underline{q}/|\underline{q}|,$$

are unit vectors.[35]

---

[35] To simplify the writing we shall omit in this appendix the explicit indications of c.m. variables, i.e., $k^* \to k$.

The conservation law $k^\mu j_\mu = 0$ allows us to work in a gauge in which the polarization vector ε_μ of the virtual photon has a vanishing time component. This can be achieved by introducing the vector

$$a_\mu = \varepsilon_\mu - k_\mu \varepsilon_0/k_0, \tag{C.2}$$

which has components

$$a_\mu \equiv (0, \underline{a}) \equiv (0, \underline{\varepsilon} - \underline{k} \frac{\underline{k}\cdot\underline{\varepsilon}}{k_0^2}). \tag{C.2'}$$

It will also be convenient to distinguish between "longitudinal" and "transverse" photons by introducing the transverse vector $\underline{b}$

$$\underline{b} = \underline{a} - \underline{k} (\underline{k}\cdot\underline{a})/\underline{k}^2 = \underline{\varepsilon} - \underline{k} (\underline{k}\cdot\underline{\varepsilon})/\underline{k}^2 \tag{C.3}$$

$$\underline{b}\cdot\underline{k} = 0. \tag{C.3'}$$

We can now represent the general electroproduction amplitude in terms of the six invariant function F_n (W, z) defined by the expansion

$$M_\mu \cdot \varepsilon^\mu = \sum_1^6 {}_n F_n (W, z) I_n, \tag{C.4}$$

where

$$\begin{aligned} I_1 &= i (\underline{\sigma}\cdot\underline{b}) & I_2 &= (\underline{\sigma}\cdot\hat{q})(\underline{\sigma}\cdot\hat{k} \times \underline{b}), \\ I_3 &= i (\underline{\sigma}\cdot\hat{k})(\underline{b}\cdot\hat{q}), & I_4 &= i (\underline{\sigma}\cdot\hat{q})(\underline{b}\cdot\hat{q}), \\ I_5 &= i (\underline{\sigma}\cdot\hat{k})(\underline{a}\cdot\hat{k}), & I_6 &= i (\underline{\sigma}\cdot\hat{q})(\underline{a}\cdot\hat{k}). \end{aligned} \tag{C.5}$$

From (C.5) it is evident that F_1, F_2, F_3, F_4 describe transitions generated by transverse photons whereas F_5, F_6 refer to transitions due to longitudinal photons.

The c.m. amplitudes F_n are related to the invariant amplitudes M_i by the rather cumbersome relations which can be found in the existing literature.

We mention that the relations among the c.m. amplitudes F_n and the invariant amplitudes M_i involve many kinematical coefficients, which make the analytic structure of the F_n rather complicated, because of the presence of the so-called "kinematical singularities". Since, on the other hand, the amplitudes F_n have simple unitarity constraints, the investigation of the combined properties of analyticity and unitarity becomes rather involved.

The "structure functions" A, B, C, D introduced in (1.13) to represent the differential cross section, namely

$$\frac{d\sigma_v}{d\Omega}(W, k^2, \varepsilon, \theta, \phi) = A + \varepsilon B + \varepsilon C \sin^2\theta \cos 2\phi + \sqrt{\varepsilon(1+\varepsilon)}\, D \sin\theta \cos\phi ,$$

can now be expressed in terms of the amplitudes F_n ($|\underline{q}| = q$, $|\underline{k}| = k$)

$$\frac{k}{q}A = \frac{1}{2}\{|F_1|^2 + |F_1 + F_4|^2 + |F_2|^2 + |F_3 + F_2|^2\} + \cos\theta\{\mathrm{Re}\, F_3^* F_4 - 2\,\mathrm{Re}\, F_1^* F_2^*\} -$$

$$- \cos^2\theta \cdot \frac{1}{2}\{|F_1 + F_4|^2 - |F_1|^2 + |F_2 + F_3|^2 - |F_2|^2\} - \cos^3\theta\, \mathrm{Re}\, F_3^* F_4,$$

$$\frac{k}{q}B = (-k^2/k_0^2)\, |F_5 + F_6|^2, \tag{C.6}$$

$$\frac{k}{q}C = \cos\theta \quad \mathrm{Re}\, F_3^* F_4 + \{|F_1 + F_4|^2 - |F_1|^2 + |F_2 + F_3|^2 - |F_2|^2\}$$

$$\frac{k}{q}D = \left\{2\,\mathrm{Re}\left[F_5^*(F_1 + F_4) + F_6^*(F_2 + F_3)\right] + 2\cos\theta\, \mathrm{Re}\,(F_3 F_5^* + F_4 F_6^*)\right\}\frac{(-k^2)^{1/2}}{k_0} .$$

We are now ready to perform the "multipole analysis" of the electroproduction amplitude, i.e., to expand the functions $F_n(W,z)$ in terms of partial amplitudes referring to the angular momentum and parity of the final state.

The angular momentum analysis of electroproduction proceeds as follows:
1) Let us consider the final state, which is composed of a spin - $\frac{1}{2}$ nucleon and of a spinless pion of odd parity. To a given value of the total angular momentum j correspond two different values of $l = j \pm \frac{1}{2}$. Since the parity of the final state is $(-1)^{l+1}$, the two channels decouple because they correspond to opposite values of the parity. One usually represents the final states by (l^+), (l^-), and the corresponding pion-nucleon phase shifts are indicated as $\delta_{l\pm}$.
2) The analysis of the initial state is more complicated because the initial virtual photon is a spin-one particle. One first defines as "multipolarity" λ the total (orbital +spin) angular momentum *carried by the initial photon*. Again for a given multipolarity, the total angular momentum j will take the value $j = \lambda \pm \frac{1}{2}$. Of course, the multipolarity λ does not completely define the state of the initial photon, because there are three independent ways of coupling the orbital angular momentum of the photon to its spin. One usually starts by separating the effects of longitudinal and transversal photon. For a given λ, a longitudinal photon carries parity $(-1)^\lambda$.

There are two kinds of transversal transitions: *electric* with parity $(-1)^\lambda$ and *magnetic* with parity $(-1)^{\lambda+1}$. In summary: in correspondence with a given multipolarity λ there are three kinds of transitions: longitudinal (L) and electric (E) with parity $(-1)^\lambda$; magnetic (M) with parity $(-1)^{\lambda+1}$.

The complete kinematics of electroproduction, taking into account parity and angular momentum conservation is shown in Table C.1.

Table C.1

Final state	Phase shift	Initial state		Amplitude
$l = j - 1/2$ $P = (-1)^{l+1} =$ $= (-1)^{j+1/2}$	δ_{l-}	$\lambda = j - 1/2$ $P = (-1)^{\lambda+1}$	magnetic	M_{l-}
		$\lambda = j + 1/2$ $P = (-1)^{\lambda}$	electric long.	E_{l-} L_{l-}
$l = j + 1/2$ $P = (-1)^{l+1} =$ $= (-1)^{j-1/2}$	δ_{l+}	$\lambda = j - 1/2$ $P = (-1)^{\lambda}$	electric long.	E_{l+} L_{l+}
		$\lambda = j + 1/2$ $P = (-1)^{\lambda+1}$	magnetic	M_{l+}

We have indicated by $M_{l\pm}(W)$, $E_{l\pm}(W)$, $L_{l\pm}(W)$ the multipole amplitudes corresponding to final states labelled by l and $j \pm \frac{1}{2}$, respectively.

Now using projection operator techniques, the invariant amplitudes $F_n\,(W,z)$ can be shown to have the following multipole expansion

$$F_1\,(W,z) = \sum_{0}^{\infty}{}_{l}\left\{\left[lM_{l+}(W) + E_{l+}(W)\right] P'_{l+1}(z) + \left[(l+1)\,M_{l-}(W) + E_{l-}(W)\right] P'_{l-1}\,(z)\right\}$$

$$F_2\,(W,z) = \sum_{1}^{\infty}{}_{l}\left[(l+1)\,M_{l+}(W) + lM_{l-}(W)\right] P'_{l}(z),$$

$$F_3\,(W,z) = \sum_{1}^{\infty}{}_{l}\left\{\left[(E_{l+}(W) - M_{l+}(W)\right] P''_{l+1}(z) + \left[E_{l-}(W) + M_{l-}(W)\right] P''_{l-1}(z)\right\},$$

$$F_4\,(W,z) = \sum_{2}^{\infty}{}_{l}\left[M_{l+}(W) - E_{l+}(W) - M_{l-}(W) - E_{l-}(W)\right] P''_{l}(z), \tag{C.7}$$

$$F_5\,(W,z) = \sum_{0}^{\infty}{}_{l}\left[(l+1)\,L_{l+}(W)\,P'_{l+1}(z) - lL_{l-}(W)\,P'_{l-1}(z)\right]$$

$$F_6\,(W,z) = \sum_{1}^{\infty}{}_{l}\left[lL_{l-}(W) - (l+1)\,L_{l+}(W)\right] P'_{l}(z).$$

From these formulae one can notice that the following multipoles never appear in the expansions and are therefore not physical:

$$E_{0-},\ E_{1-},\ M_{0+},\ M_{0-},\ L_{0-}. \tag{C.8}$$

According to the Fermi-Watson theorem $M_{l\pm}$, $E_{l\pm}$, $L_{l\pm}$ are complex amplitudes whose phase is given by the corresponding pion-nucleon phase shift $\delta_{l\pm}$. As a consequence of the fact that at low energy the important pion-nucleon phase shift is the j = 3/2, I = 3/2 one, the only multipole amplitudes with a large imaginary part are M_{1+}, E_{1+}, L_{1+}. This fact is at the root of all isobaric and dispersion models.

The inverse relation between the multipoles and the c.m. amplitudes F_i or the invariant amplitudes M_i involves a projection by angular integration and is a complicated one. Since our considerations in the text refer mainly to the threshold configuration q = 0, l = 0 [see (C.12)], we reproduce here the expression of E_{0+}, L_{0+} at threshold in terms of the invariant amplitudes M_i of (3.15). One has

$$E_{0+}\Big|_{th.} = -\sqrt{P^2/m_N^2}\,\left(M_1 + \sqrt{m_N^2/P^2}\,\nu\, M_5\right)\Big|_{th.} \,, \qquad (C.9)$$

$$\frac{L_{0+}}{k_0}\Big|_{th.} = \frac{1}{m_N(2m_N^2+m_\pi)}\sqrt{m_N^2/P^2}\,\Big\{ m_\pi \nu\left[M_6 + \frac{\nu}{P^2}(M_7+M_8) - \frac{m_N}{P^2}M_5\right] - $$
$$- \nu\left[M_2 + \frac{\nu}{P^2}(M_3+M_4) - 2\,m_N\left(M_6 + \frac{\nu}{P^2}M_7 + \frac{\nu}{P^2}M_8\right) + 2\,M_5\left(1 - \frac{q\cdot\Delta}{4P^2}\right)\right]\Big\}\Big|_{th.} \,, \qquad (C.10)$$

with $P^2 = m_N^2\,(1-t/4m_N^2)$, and all kinematical variables must be taken at their threshold values.

An important property of the multipole amplitudes is represented by their threshold behaviour, which can be derived, for instance, from elementary analyticity requirements. One has that

$$\begin{aligned} &E_{l+} \sim k^l\, q^l, \quad && L_{l+} \sim k^l\, q^l \quad && l \geq 0, \\ &M_{l+} \sim k^l\, q^l \quad && M_{l-} \sim k^l\, q^l \quad && l \geq 1, \\ &E_{l-} \sim k^{l-2}\, q^l, \quad && L_{l-} \sim k^{l-2}\, q^l \quad && l \geq 2, \end{aligned} \qquad (C.11)$$

(except $L_{1-} \sim kq$).

As a consequence of (C.11) it is easy to obtain that

$$\frac{k}{q}A \underset{q\to 0}{\longrightarrow} |E_{0+}|^2_{th.}\,, \qquad \frac{k}{q}B \underset{q\to 0}{\longrightarrow} \left(-\frac{k^2}{k_0^2}\right)|L_{0+}|^2_{th.}\,, \qquad C,\, D \underset{q\to 0}{\longrightarrow} 0. \qquad (C.12)$$

A further interesting fact is the proportionality between longitudinal and electric matrix elements in the unphysical point k = 0 (in the c.m.). Such a relationship can be shown to follow from the current conservation requirement and takes on the form

$$\lim_{k \to 0} E_{l+}/L_{l+} = 1, \qquad \lim_{k \to 0} E_{l-}/L_{l-} = -l/(l-1). \tag{C.13}$$

The reader can easily check the validity of conditions (C.13) in the particular threshold configuration $l = 0$. It is enough to use the expressions (C.9), (C.10) in terms of the invariant amplitudes. The point $q = 0$ $k = 0$ corresponds to $\nu = m_N m_\pi$ $t = 0$ $k^2 = m_\pi^2$, and using the gauge invariance constraints (3.24) one ascertains at once that $E_{0+} = L_{0+}$.

To conclude this appendix it may be useful to reproduce the multipole expansions of the transversal and longitudinal cross section which result after integration of (1.13) over the pion angles. These are

$$\begin{aligned} \sigma_T &= 2\pi \frac{q}{k} \sum_{0}^{\infty} {}_l \left\{ l(l+1)\left[|M_{(l+1)+}|^2 + |E_{(l+1)-}|^2\right] + l^2 (l+1)\left[|M_{l-}|^2 \; k \; |E_{(l-1)+}|^2\right]\right\} \\ \sigma_L &= 4\pi \frac{q}{k} \sum_{0}^{\infty} {}_l \left[(l+1)^3 \, |L_{(l+1)-}|^2 + l^3 |L_{(l-1)+}|^2\right]. \end{aligned} \tag{C.14}$$

Again one finds at threshold that

$$\begin{aligned} \sigma_T &\to 4\pi \frac{q}{k} \, |E_{0+}|^2_{th.} \; , \\ \sigma_L &\to 4\pi \frac{q}{k} \, (-k^2) \left|\frac{L_{0+}}{k_0}\right|^2_{th.} \; . \end{aligned} \tag{C.15}$$

An alternative decomposition can be used, which corresponds to a virtual photon which transverse and scalar components, $\underline{b}$ and b_0 respectively ($b_0 = \varepsilon_0 - k_0 \frac{k \cdot \varepsilon}{k^2}$). Then a scalar cross section σ_S and scalar multipoles $S_{l\pm}$ are introduced instead of the longitudinal quantities. Sometimes these quantities are used in the text.

Appendix D: Regge Behavior and Gauge Invariance

1) Establishing the Regge behaviour and the nature of the trajectories contributing for each invariant amplitude $M_i(\nu,t)$ is a well-known and solved problem whose details can be found in the literature /213/. For our purposes Table D.1 is sufficient, where we indicate also the relevant leading trajectories. We have included among them, beside the familiar entities π, ρ, ω, the $B(J^P = 1^+$, $M = 1228$ MeV) and the $A_2(J^P = 2^+$, $M = 1310$ MeV) resonances and the $(\rho\pi)$ partial wave enhancement $A_1(J^P=1^+$, $M \simeq 1100$ MeV) /214/.

Table D.1.

	(-) I = 1, G = -1	(+) I = 0, G = -1	(o) I = 1, G = 1
M_1	$\nu^{\alpha_{A_1}-1}$, $\nu^{\alpha_{A_2}}$	ν^{α_ω}	ν^{α_ρ}
M_2	$\nu^{\alpha_\pi-1}$, $\nu^{\alpha_{A_2}-1}$ $\nu^{\alpha_{A_1}-2}$	ν^{α_ω}	$\nu^{\alpha_\rho-1}$, ν^{α_B-1}
$M_{3,4}$	ν^{α_π}, $\nu^{\alpha_{A_2}}$, $\nu^{\alpha_{A_1}-1}$	ν^{α_ω}	ν^{α_ρ} ν^{α_B}
M_5	$\nu^{\alpha_{A_2}-1}$	$\nu^{\alpha_\omega-1}$	$\nu^{\alpha_\rho-1}$
M_6	$\nu^{\alpha_{A_2}-1}$, $\nu^{\alpha_{A_1}-2}$	$\nu^{\alpha_\omega-1}$	$\nu^{\alpha_\rho-1}$
$M_{7,8}$	$\nu^{\alpha_{A_2}-2}$, $\nu^{\alpha_{A_1}-1}$	$\nu^{\alpha_\omega-2}$	$\nu^{\alpha_\rho-2}$

We can exploit these results to illustrate the mechanism which generates singularities in the t-channel, which correspond to the exchange of physical states of given mass and spin. We concentrate on the pion case.

It turns out (see Table D.1) that the pion trajectory contributes to the amplitudes $M^{(-)}_{2,3,4}$ and, without taking into account multiplying factors, their asymptotic form is

$$M^{(-)}_{2} \underset{\nu\to\infty}{\sim} \beta^{(2)}_{\pi}\ (t,\ k^2)\ \alpha_{\pi}(t)\nu^{\alpha_{\pi}(t)-1}, \qquad \text{(D.1)}$$

$$\text{(D.1)}$$

$$M^{(-)}_{3,4} \underset{\nu\to\infty}{\sim} \beta^{(3,4)}_{\pi}\ (t,\ k^2)\ \nu^{\alpha_{\pi}(t)}.$$

A glance at the dispersion relation shows that, while for the antisymmetric amplitude $M^{(-)}_{2}$ there is no need of subtractions, the situation for $M^{(-)}_{3,4}$ requires some care. Indeed

$$M^{(-)}_{3,4}\ (\nu,\ t,\ k^2) = M^{(-)}_{3,4}\Big|_{\text{nucleon}} + \frac{2}{\pi}\int_{\nu_0}^{\infty}\frac{\nu' d\nu'}{\nu'^2-\nu^2}\ \text{Im}\ M^{(-)}_{3,4}\ (\nu',\ t,\ k^2), \qquad \text{(D.2)}$$

and an asymptotic damping is required for the convergence of the integral. This clearly depends on the behaviour of $\alpha_{\pi}(t)$ for negative t. Since $\alpha_{\pi}(m^2_{\pi}) = 0$ one could reasonably expect $\alpha_{\pi}(t) < 0$ for not too large spacelike t, so that no subtraction needs be performed. The situation, however, becomes delicate as $t \to m^2_{\pi}$, $\alpha_{\pi}(t) \to 0$. To see this more explicity, we isolate the asymptotic region of integration by introducing a cut-off Λ. Then one has approximately, for $t < 0$,

$$\beta_{\pi}(t)\int_{\Lambda}^{\infty}\frac{d\nu'}{\nu'}\ (\nu')^{\alpha_{\pi}(t)} \sim \frac{1}{\alpha_{\pi}(t)}\ (\Lambda)^{\alpha_{\pi}(t)}\ \ \beta_{\pi}(t),$$

and, as $\alpha_{\pi}(t) \to 0$, a pole develops, namely

$$\frac{1}{\alpha_{\pi}(t)} \sim \frac{1}{\alpha'_{\pi}(m^2_{\pi})(t-m^2_{\pi})}\ . \qquad \text{(D.3)}$$

Thus, by suitable normalizing the residue functions $\beta^{(3,4)}_{\pi}\ (t = m^2_{\pi},\ k^2)$, the elementary pion singularity is reproduced without need of introducing it from the beginning. (The same argument can, of course, be applied to other trajectories.) This corresponds to the idea that in the framework of Regge pole theory there are no elementary particles but they all lie on trajectories.

Finally, coming back to the general problem of subtractions, one can easily verify, combining the results of Table D.1 with the current phenomenological indication $\alpha(t) < 1$ for small spacelike t [36], that no subtractions are required in the framework of

---

[36] On general grounds $\alpha(0) \leq 1$ and experimentally /215/

$\alpha_{\rho}(0) \simeq \alpha_{\omega}(0) \simeq 0.5$, $\alpha_{A_2}(0) \simeq 0.4$.

the simple Regge pole model. In the evaluation of electroproduction amplitudes by dispersion relations, experts usually do adopt a one-subtracted dispersion relation for $M_{3,4}^{(-)}$ (even if in principle not necessary) to improve the slow convergence of the integral.

2) An interesting point arises when one combines the requirements coming from the gauge invariance constraint [(3.24)] and from the assumptions of analyticity and Regge behaviour of the amplitudes.

We first of all notice that when $k^2 \neq 0$, it is not possible to obtain simple and general statements such as the Kroll-Ruderman theorem for (soft-pion) photoproduction. Also the role of kinematical singularities is a little more delicate since, for instance, the point $q \cdot k = 0$ is now in the physical region [37].

Thus one must be certain that, for the sake of writing a gauge invariant decomposition, no unwanted singularities affect the physical matrix element. This gives rise to particular constraints among the amplitudes which, when expressed through the dispersion relations, lead to sum rules.

As an instructive example we proceed to the derivation of a relation between the form factors $F_1^V(k^2)$ and $F_\pi(k^2)$, which can be considered as the natural generalization of the universality condition (3.27). Indeed, take the limit $\nu \to 0$ at fixed $q \cdot k$, k^2 in the first of (3.24), express the invariant amplitudes via unsubtracted dispersion relations and select nonvanishing polar contributions; the result is

$$g_{\pi N} F_1^{(v)}(k^2) = -\frac{2}{\pi}\int_{\nu_0}^{\infty} \frac{d\nu'}{\nu'} [q \cdot k \text{ Im } M_3^{(-)}(\nu',t,k^2) + k^2 \text{ Im } M_4^{(-)}(\nu',t,k^2)]. \qquad (D.4)$$

If one goes to the limit $t \to m_\pi^2$, the pion trajectory must again be selected from the high-energy tail of the dispersive integral. Since the residues of the pion trajectory are related to the electromagnetic pion form factor, the final result turns out to be

$$g_{\pi N} [F_1^{(v)}(k^2) - F_\pi(k^2)] = -\frac{2k^2}{\pi} \cdot \int_{\nu_0}^{\infty} \frac{d\nu'}{\nu'} \text{ Im } [\tilde{M}_4^{(-)}(\nu',t=m_\pi^2,k^2) + \frac{1}{2}\tilde{M}_3^{(-)}(\nu',t=m_\pi^2,k^2)] \qquad (D.5)$$

(the ~ symbol indicates that the pion trajectory has already been selected).

An indicative estimate of the continuum integral in (D.5) provides for $\langle r_\pi^2 \rangle^{1/2}$ a value not very different from the corresponding one for the nucleon, i.e., $\langle r_\pi^2 \rangle^{1/2} \approx$ 0.76 Fermi /216/.

---

[37] The physical region for t is $t < (k^2 - m_\pi^2)(1 + m_\pi/m_N)^{-1}$. Then $q \cdot k = 0$ requires $t = m_\pi^2 + k^2$, which is still physical provided $k^2 < -m_\pi m_\pi (2 + m_\pi/m_N)$.

The existence of a relation between $F_1^{(v)}(k^2)$ and $F_\pi(k^2)$ is not suprising, by the way, if one remembers that the electroproduction generalized Born approximation, i.e., the sum of all polar contributions, is not gauge invariant by itself, so that additional terms [the continuum integral of (D.5)] must be present to restore gauge invariance. For practical purposes, however, phenomenological descriptions are often used with gauge invariant Born approximations, built ad hoc, which embody the pion pole term /217/.

References

1 W.K.H. Panofsky, E.A. Allton: Phys. Rev. 110, 1155 (1958)
2 E. Amaldi: "The Electroproduction of Pions"; in *Frascati Meeting on the Electrosynchrotron*, Frascati, 5-7 November 1970; "Electroproduction of Pions at Low Energy"; in *Topical Seminar on Electromagnetic Interactions*, Trieste, 21-26 June (1971)
3 A.B. Clegg: "Electroproduction in the Resonance Region"; in *Int. Symp. on Electron and Photon Interactions at High Energies*, Bonn, August 1973 (North-Holland, Amsterdam 1974)
4 M. Gourdin: "Weak and Electromagnetic Form Factors of Hadrons"; Physics Report 11C, 29 (1974)
5 J. Gayler: "Electroproduction in the Resonance Region", VIII All Soviet Union High Energy Physics School, Erevan, April 1975
6 G. Wolf: "Review of Electroproduction of Final States"; in *Int. Symp. on Lepton and Photon Interactions at High Energies* (Stanford University, Stanford, August 1975)
7 A. Donnachie, G. Shaw, D. Lyth: *Form Factors and Electroproduction* (University of Manchester, Preprint, 1977)
8 J.S. O'Connell, P.A. Tipler, P. Axel: Phys. Rev. 126, 228 (1962)
9 See for example: J. Ballam, G.B. Chadwick, Z.D.T. Guiragossian, A. Kilert, R.R. Larsen, D.W.G.S. Leith: Nucl. Instr. Methods 73, 53 (1969)
10 D.I. Drickey, R.F. Mozley: Phys. Rev. Lett. 8, 291 (1962)
11 N. Cabibbo, G. Da Prato, G. De Franceschi, U. Mosco: Phys. Rev. Lett. 9, 270, 435 (1962)
C. Berger, G. Mc Clellan, N. Mistry, H. Ogren, B. Sandler, J. Swartz, P. Walstrom, R.L. Anderson, D. Gustavson, J. Johnson, I. Overman, R. Talman, B.H. Wick, D. Worcester, A. Moore: Phys. Rev. Lett. 25, 1366 (1970)
12 G. Diambrini Palazzi: Rev. Mod. Phys. 40, 611 (1968)
13 R.H. Milburn: Phys. Rev. Lett. 10, 75 (1963)
F.R. Arutyunian, I.I. Goldman, V.A. Tumanian: Soviet Phys.-JETP 18, 218 (1964)
14 C.W. Akerlof, W.W. Ash, K. Berkelman, C.A. Lichtenstein: Phys. Rev. Lett. 16, 147 (1960)
C.W. Akerlof, W.W. Ash, K. Berkelman, C.A. Lichtenstein, A. Ramanauskas, R.H. Siemanrn: Phys. Rev. 163, 1482 (1967)
H.F. Jones: Nuovo Cimento 40A, 1018 (1965)
15 For a recent description see, for instance, V. De Alfaro, S. Fubini, G. Furlan, C. Rossetti: *Currents in Hadron Physics* (North-Holland, Amsterdam, 1973)
16 L.N. Hand: Phys. Rev. 129, 1834 (1963)
17 See for example: S.D. Drell, F. Zachariasen: *Electromagnetic Structure of Nucleons* (Oxford University Press, Oxford, 1961)
18 For a recent review which also contains a discussion of theoretical models, see G. Höhler "Dispersion theory of nucleon form factors", Lecture Notes in Physics 56, 159 (1976), Springer-Verlag. An updated review of the experimental situation can be found in G. Höhler, E. Pietarinen, I. Sabba-Stefanescu, F. Borkowski, G.G. Simon, V.H. Walter and R.D. Wendling: Nucl. Phys. B114, 505 (1976)
19 D.H. Wilkinson, D. Alburger: Phys. Rev. Lett. 26, 1127 (1971)
See, however, for a recent possible positive indication: F.P. Calaprice, S.J. Freedman, W.C. Mead, H.C. Vantine: Phys. Rev. Lett. 35, 1566 (1975)

20 See the "Compilation of Coupling Constants and Low Energy Parameters" by M.M. Nagels, J.J. De Swart, H. Nielsen, G.C. Oades, J.L. Petersen, B. Tromborg, G. Gustafson, A.C. Irving, C. Jarlskog, W. Pfeil, H. Pilkuhn, F. Steiner and L. Tauscher: Nucl. Phys. B109, 1 (1976)
21 R.C.E. Devenish, R.S. Eisenschitz, J.G. Körner: Phys. Rev. D14, 3063 (1977)
J.G. Körner, M. Kuroda: Phys. Rev. D16, 2165 (1977)
22 M. Scadron, H.F. Jones: Ann. Phys. (N.Y.) 81, 1 (1973)
23 See for instance S.L. Adler: Ann. Phys. (N.Y.) 50, 189 (1968)
24 C.H. Llewellyn-Smith: Physics Report 3C, 261 (1972)
25 See for example R. Hofstadter: "Quantum Electrodynamics on Electron-Positron System"; in *Int. Symp. on Lepton and Photon Interactions at High Energies*, Stanford University, Stanford, August 1975
E. Picasso: " Recent Experimental Tests on Quantum Electrodynamics at Low Energies"; in *Abhandlungen der Deutschen Akademie der Naturforscher Leopoldina*, Supplementum Nr. 8, Bd. 44 (1976)
F. Combley, E. Picasso: "Some Topics in Quantum Electrodynamics", Course on Metrology and Fundamental Constants of the Intern. Summer School of Physics Enrico Fermi, Varenna, July (1976)
26 B. Bartolini, F. Felicetti, V. Silvestrini: La Rivista del Nuovo Cimento 2, 241 (1972), see also /18/
27 D.H. Perkins: "Review of Neutrino Experiments"; in *Int. Symp. on Electron and Photon Interactions at High Energies* (Stanford University, Stanford, August 1975)
28 R.F. Schwitters and K. Strauch: "The Physics of e^+e^- collisions", Ann. Rev. Nucl. Sci. 26, 89 (1976)
29 A. Zichichi: "Why e^+e^- Physics is fascinating", Riv. Nuovo Cimento 4, 498 (1974)
30 N. Cabibbo, G. Gatto: Phys. Rev. 124, 1577 (1961)
31 V.L. Auslander, G.I. Budker, Ju.N. Pestov, V.A. Sidorov, A.N. Skrinsky, A.G. Khabakhpashev: Phys. Lett. 25B, 433 (1967)
V.L. Auslander, G.I. Budker, E.V. Pakhtusova, Ju.N. Pestov, V.A. Sidorov, A.N. Skrinskii, A.G. Khabakhpashev: Soviet J. Nucl. Phys. 9, 69 (1969)
V.A. Sidorov: "Results of Experiments at VEPP-2M"; in *Int. Conf. on High Energy Physics*; Tbilisi, 15-21 July (1976)
32 J.E. Augustin, J.C. Bizot, J. Buon, J. Haissinski, D. Lalanne, P.C. Marin, J. Perez-y-Jorba, F. Rumpf, E. Silva, S. Tavernier: Phys. Rev. Lett. 20, 126 (1968)
J.E. Augustin, J.C. Bizot, J.C. Buon, J. Haissinski, D. Lalanne, P. Marin, H. Nguyen Ngoc, J. Perez-y-Jorba, F. Rumpf, E. Silva, A. Tavernier: Phys. Lett. 28B, 508 (1969)
D. Benaksas, G. Cosme, B. Jean-Marie, S. Jullian, F. Laplanche, J. Lefrancois, A.D. Liberman, G. Parrour, J.P. Repellin, G. Sauvage: Phys. Lett. 39B, 289 (1972)
A. Quenzer, F. Rumpf, J.L. Bertrand, J.C. Bizot, R.L. Chase, A. Cordier, B. Delcourt, P. Eschstruth, E. Fulda, G. Grosdidier, J. Hassinski, J. Jeanjean, M. Jeanjean, R. Madares, J.L. Masnou, J. Perez-y-Jorba: "Pion Form Factor in the Time-Like Region from $e^+e^-\to\pi^+\pi^-$ near Threshold", preprint LAL 1282, October (1975), and contributed paper to the "Int. Symp. on Lepton and Photon Interactions at High Energies" (Stanford 1975)
G. Cosme, A. Courau, B. Dudelzak, B. Grelaud, B. Jean-Marie, S. Jullian, D. Lalanne, F. Laplanche, G. Parroux, R. Riskalla, Ph. Roy, G. Sklarz: Int. Rep. du Lab. de L'Accelerateur Lineaire, Université Paris Sud, LAL 1287, Juillett (1976)
33 G. Barbiellini, M. Grilli, E. Iarocci, P. Spillantini, V. Valente, R. Visentin, F. Ceradini, M. Conversi, S. D'Angelo, G. Giannoli, L. Paoluzzi, R. Santonico: Lettere al Nuovo Cimento 6, 557 (1973)
34 V. Alles Borelli, M. Bernardini, D. Bollini, P.L. Brunini, E. Fiorentino, T. Massam, L. Monari, F. Palmonari, F. Rimondi, A. Zichichi: Phys. Lett. 40B, 433 (1972)
M. Bernardini, D. Bollini, P.L. Brunini, E. Fiorentino, T. Massam, L. Monari, F. Palmonari, F. Rimondi, A. Zichichi: Phys. Lett. 44B, 393 (1973)
D. Bollini, P. Giusti, T. Massam, L. Monari, F. Palmonari, G. Valenti, A. Zichichi: Lettere al Nuovo Cimento 14, 418 (1975)
M. Basile, D. Bollini, G. Cararomeo, L. Ciafarelli, P. Giusti, T. Massam, L. Monari, F. Palmonari, M. Placidi, G. Valenti, A. Zichichi: Nuovo Cimento 34A, 1 (1976)

35 J. Allan, G. Ekspong, P. Sällström, K. Fisher: Nuovo Cimento 32, 1144 (1964)
36 D.G. Cassel: Thesis (1965) unpublished; P. Shepard, APS Meeting, April (1972)
37 G.T. Adylov, F.K. Aliev, D. Yu. Bardin, W. Gajewski, I. Ion, B.A. Kulakov, G.V. Michelmacher, B. Niezyporuk, T.S. Nigmanov, E.N. Tsyganov, M. Turala, A.S. Vodopianov, K. Wala, E. Dally, D. Drickey, A. Liberman, P. Shepard, J. Thompkins, C. Buchanan, J. Poisier; Phys. Lett. 51B, 402 (1974)
38 E.B. Dally, D.J. Drickey, J.M. Hauptman, C.F. May, D.H. Stork, J.A. Poirier, C.A. Rey, R.J. Wojslaw, P.F. Shepard, A.J. Lennox, J.C. Tompkins, T.E. Toohig, A.A. Wehmann, I.X. Joan, T.A. Nigmanov, E.N. Tsyganov and A.S. Vodopianov: Phys. Rev. Lett. 39, 1176 (1977)
39 M.M. Sternheim, R. Hofstadter: Nuovo Cimento 38, 1854 (1965)
40 P. Demeur: Phys. Rev. 124, 2000 (1961)
M. Ericson: Nuovo Cimento 32, 251 (1964)
H.E. Nordberg, K.F. Kinsey: Phys. Rev. Lett. 20, 692 (1966)
41 M.M. Block, I. Kenyon, J. Keren, D. Koetke, P. Malhotra, R. Walker, H. Winzeler: Phys. Rev. 169, 1074 (1968)
42 K.M. Crowe, A. Fainberg, J. Miller, A.S.L. Parsons: Phys. Rev. 180, 1349 (1969)
43 R.A. Christensen: Phys. Rev. D1, 1469 (1970)
G.C. Oades, G. Rasche: Nucl. Phys. B20, 333 (1970)
C.T. Mottershead: Phys. Rev. D6, 780 (1972)
44 F. Nichitiu: *Proc. Fifth Int. Conf. High Energy Physics and Nuclear Structure*, Uppsala, 28-22 Juni (1973) ed. by G. Tibell (North-Holland, Amsterdam 1974, and Dubna report E2-6890, 1973) p. 178
45 F. Nichitiu, Yu.A. Shcherbakov: Nucl. Phys. B61, 429 (1973)
46 Yu.A. Shcherbakov, T. Angelescu, I.V. Falomkin, M.M. Kulyukin, V.I. Lyashenko, R. Mach, M. Mihul, Nguyen Minh Kao, F. Nichitiu, G.P. Pontecorvo, V.K. Sarychieva, M.G. Sapozhnikov, M. Semerdjieva, T.M. Troshec, N.I. Trosheva, F. Balestra, L. Busso, R. Garfagnini, G. Piragino: Nuovo Cimento A17, 355 (1973)
47 C. Mistretta, D. Imrie, J.A. Appel, R. Budnitz, L. Carroll, J. Chen, J. Dunning, M. Goitein, K. Hanson, A. Litke, R. Wilson: Phys. Rev. Lett. 20, 1070 (1968)
D. Imrie, C. Mistretta, R. Wilson: Phys. Rev. Lett. 20, 1074 (1968); Phys. Rev. 184, 1487 (1969); D3, 2923 (1971)
C. Mistretta, D. Imrie, J.A. Appel, R. Budnitz, L. Carroll, M. Goitein, K. Hanson, R. Wilson: Phys. Rev. Lett. 20, 1523 (1968)
48 C.J. Bebeck, C.N. Brown, M. Herzlinger, S.D. Homes, C.A. Lichtenstein, F.M. Pipkin, S. Raither, L.K. Sisterson: Phys. Rev. D13, 25 (1976)
49 S. Devons, E. Di Capua, A. Lanzara, P. Nemethy, C. Nissim-Sabat: Phys. Rev. 184, 1356 (1969)
50 S.F. Berezhnev, A.V. Demyanov, A.V. Kulikov, A.V. Kuptsov, V.P. Kurochkin, G.G. Mkrtchyan, L.L. Nemenov, Zh.P. Pustyl'nik, G.I. Smirnov, A.G. Fedunov, D.M. Khazins: Soviet J. Nucl. Phys. 18, 53 (1974)
51 S.J. Brodsky, G.R. Farrar: Phys. Rev. D11,1309 (1975)
52 G. Di Giugno, J.W. Humphrey, E. Sassi, G. Troise, U. Troya, S. Vitale, M. Castellano: Lettere al Nuovo Cimento 2, 873 (1971)
53 M. Conversi, T. Massam, Th. Muller, A. Zichichi: Nuovo Cimento 40A, 690 (1965). For a fairly recent measurement see G. Bassompierre, G. Binder, P. Dalpiaz, P.F. Dalpiaz, G. Gissinger, S. Jacquey, C. Peroni, M.A. Schneegans, L. Tecchio: Phys. Lett.68B, 477 (1977)
54 D.L. Hartill, B.C. Barish, D.G. Fong, R. Gomez, J. Pine, A.V. Tollestrup: Phys. Rev. 184, 1485 (1969)
55 L. Koester, W. Nistler, W. Waschkowski: Phys. Rev. Lett. 36, 1021 (1976). See also L. Koester, Springer Tracts in Modern Physics 80 (1977)
56 S.J. Barish, J. Campbell, G. Charlton, Y. Cho, M. Derrick, R. Engelmann, L.G. Hyman, D. Jankowski, W.A. Mann, B. Musgrave, P. Schreiner, P.F. Schultz, R. Singer, M. Szczekowski, T. Wangler, H. Yuta, V. Barnes, D. Carmany, A. Garfinkel, G. Radecky: "Study of Neutrino Interactions in Hydrogen and Deuterium I." Argonne National Laboratory, June (1977)
See also M. Derrick: "Quasi-elastic Reactions"; in *Int. Symp. on Electron and Photon Interactions at High Energies*, Bonn, August 1973 (North-Holland, Amsterdam, 1974)
57 See N.C. Mukhopadhyay: Physics Report 30C, 1 (1977)

58 P. Schreiner, E. von Hippel: Nucl. Phys. B58, 333 (1973)
59 See for instance D. Amati, S. Fubini: Ann. Rev. Nucl. Sci. 12, 359 (1962)
R.J. Eden: *High Energy Collisions of Elementary Particles* (Cambridge University Press, Cambridge, 1967)
B.H. Bransden, R.G. Moorhouse: *The Pion-Nucleon System* (Princeton University Press, Princeton, 1973)
60 See for instance: V. De Alfaro, S. Fubini, G. Furlan, C. Rossetti: *Currents in Hadron Physics* (North-Holland, Amsterdam, 1973)
S.L. Adler, R. Dashen: *Current Algebras* (W.A. Benjamin, 1968)
61 See for instance: P. Dennery: Phys. Rev. 41, 236 (1969)
S.L. Adler: Ann. Phys. (N.Y.) 50, 189 (1968)
A. Donnachie: *Photo- and Electroproduction Processes in High Energy Physics*, Vol. 5, ed. by E.H.S. Burhop, (Academic Press, New York, 1972)
G. von Gehlen: "Theoretical Aspects of Photo- and Electroproduction in the Resonance Region", in *Symp. on Electron and Photon Interactions at High Energies*, Bonn, August 1973 (North-Holland, Amsterdam, 1974). These references contain a fairly complete bibliography on the kinematical and dynamical aspects of the electroproduction process.
62 J.S. Ball: Phys. Rev. 124, 2014 (1961)
63 N. Dombey: Nuovo Cimento 32, 1969 (1964)
64 N.M. Kroll, M.A. Ruderman: Phys. Rev. 93, 233 (1954)
65 G. Cochard, M. Karatchentzeff, P. Kessler, B. Roehner: Nuovo Cimento A12, 909 (1972). This paper contains reference to the previous works on the subject
66 G. von Gehlen: Nucl. Phys. B9, 17 (1969); B20, 102 (1970)
67 G. von Gehlen: "Threshold Pion Electro- and Photoproduction", Bonn University Report 2-80 (1970)
G. von Gehlen, H. Wessel: "Electroproduction Coincidence Cross Sections from Coupled Multipole Equations", Bonn University Report 2-94 (1971)
68 See for a review: A.C. Hearn, S.D. Drell: In High Energy Physics, Vol. 2, ed. by E.H.S. Burhop (Academic Press, New York, 1967)
69 J. Bernstein, S. Fubini, M. Gell-Mann, W. Thirring: Nuovo Cimento 17, 79 (1960)
Chou Kuang Chao: Soviet Phys.-JETP 12, 492 (1961). For a general review of the topics discussed in this section see /2/
70 M. Gell-Mann, M. Levy: Nuovo Cimento 16, 705 (1958)
71 M.L. Goldberger, S.B. Treiman: Phys. Rev. 110, 1178 (1958)
72 S.L. Adler: Phys. Rev. 137B, 1022 (1965)
73 M. Gell-Mann: Phys. Rev. 125, 1067 (1962); Physics 1, 63 (1964)
74 Y. Tomozawa: Nuovo Cimento 46A, 707 (1966)
S. Weinberg: Phys. Rev. Lett. 17, 616 (1966)
75 See /73/ and N. Cabibbo: Phys. Rev. Lett. 10. 531 (1963)
76 S.S. Gershtein, I.B. Zeldovich: Soviet Phys.-JETP 2, 576 (1956)
R.P. Feynman, M. Gell-Mann: Phys. Rev. 109, 193 (1958)
77 Y. Nambu: Phys. Rev. Lett. 4, 380 (1960)
78 S. Fubini, G. Furlan: Ann. Phys. (N.Y.) 48, 322 (1968)
G. Furlan, N. Paver, C. Verzegnassi: Nuovo Cimento 62A, 519 (1969)
79 G. Furlan, N. Paver, C. Verzegnassi: Nuovo Cimento 20A, 295 (1974)
80 See among the most recent papers: J.F. Gunion, P.C. Mc Namee, M.D. Scadron: Nucl. Phys. B123, 445 (1977), S. Weinberg: *The problem of mass*, Harvard Univ. preprint HUTP-77/A057 (1977)
J.F. Gunion, P.C. Mc Namee, M.D. Scadron: Nucl. Phys. B123, 445 (1977)
81 Y. Nambu, D. Lurie: Phys. Rev. 125, 1429 (1962)
Y. Nambu, E. Shrauner: Phys. Rev. 128, 862 (1962)
82 see: C. Verzegnassi: Daresbury Lecture Note Series DNPL/R8
83 Y. Nambu, M. Yoshimura: Phys. Rev. Lett. 24, 25 (1970)
84 G. Benfatto, F. Nicolo, G.C. Rossi: Nucl. Phys. B50, 205 (1972); Nuovo Cimento 14A, 425 (1973)
85 S.L. Adler, F. Gilman: Phys. Rev. 152, 1460 (1966)
86 N. Dombey, B.J. Read: Nucl. Phys. B60. 65 (1973)
A.I. Vainshtein, V. Zakharov: Nucl. Phys. B36, 589 (1972)
87 G. Furlan, N. Paver, C. Verzegnassi: Nuovo Cimento 70A, 247 (1970); Springer Tracts in Modern Physics 62, 118 (1972)

88 G. van Gehlen, M.G. Schmidt: Nucl. Phys. B20, 173 (1970); R. Mc. Claskey, R.Y. Jacob, G.E. Hite: "Applications of interior dispersion relations to pion photoproduction near threshold". Arizona State University (Tempe) preprint, ASU-HEP 10177 (1977)
89 J.P. Burq: Ann. Phys. (Paris) 10, 363 (1965)
90 B.B. Govorkov, S.P. Denisov, E.V.J. Minarik: Soviet J. Nucl. Phys. 4, 265 (1966)
91 B.B. Govorkov, S.P. Denisov, E.V. Minarik; see /29/ and Soviet J. Nucl. Phys. 4, 265 (1967)
but see also: P. Noelle: Bonn University preprint PI 2-92 (1971)
92 For the sum rule (3.75): S. Fubini, G. Furlan, C. Rossetti: Nuovo Cimento 40A, 1171 (1965)
for the sum rule (3.74): S.L. Adler; *Proc. Int. Conf. on Weak Interactions*, Argonne, 1965
G. Furlan, R. Jengo, E. Remiddi: Nuovo Cimento 44A, 427 (1966)
Riazuddin, B.W. Lee: Phys. Rev. 146B, 1202 (1966)
93 See /85/ and G. von Gehlen, M.G. Schmidt: Nucl. Phys. B20, 173 (1970)
R.C.E. Devenish, D.H. Lyth: Nucl. Phys. B43, 228 (1972)
94 R.A. Berg, C.N. Linder: Nucl. Phys. 26, 259 (1961); Phys. Rev. 112, 2072 (1958)
P.S. Isaev, I.S. Zlatev: Nucl. Phys. 16, 608 (1960); Nuovo Cimento 13, 1 (1959)
M. Greco, A. Verganelakis: Phys. Lett. 27B, 317 (1968)
95 C.J. Bebek, C.N. Brown, M. Herzlinger, S.D. Holmes, C.A. Lichtenstein, F.M. Pipkin, L.K. Sisterson, D. Andrews, K. Berkelman, D.G. Cassel, D.L. Hartill: Phys. Rev. D9, 1229 (1974)
96 A. Del Guerra, A. Giazotto, M.A. Girogi, A. Stefanini, D.R. Botterill, D.W. Braben, D. Clarke, P.R. Norton: Nucl. Phys. B99, 253 (1975)
97 W. Bartel, F.W. Büsser, W.R. Dix, R. Felst, D. Harms, H. Krehbiel, P.E. Kuhlmann, J. Mc. Elroy, J. Meyer, G. Weber: Nucl. Phys. B58, 429 (1973)
98 P. Brauel, F.W. Büsser, Th. Canzler, D. Cords, W.R. Dix., R. Felst, G. Grundhammer, W.D. Kollmann, H. Krehbiel, J. Meyer, G. Weber: Phys. Lett 45B, 389 (1973)
99 N. Meister, D.R. Yennie: Phys. Rev. 130, 1210 (1963)
100 A. Bartl, P. Urban: Acta Phys. Austriaca 24, 139 (1966)
101 L.W. Mo, Y.S. Tsai: SLAC-Pub. 830 (1968); Rev. Mod. Phys. 41, 205 (1969)
102 R.D. Kohaupt: Z. Physik 194, 18 (1966)
103 Y.S. Tsai: SLAC-Pub. 848 (1971)
104 P. Urban: *Topics in Applied Quantum-Electrodynamics* (Springer Verlag, Wien - New York 1970)
105 F.J. Gilman: Phys. Rev. 167, 1365 (1967)
106 H.L. Lynch, J.V. Allaby, D.M. Ritson: Phys. Rev. 164, 1635 (1967)
107 F.W. Brasse, J. Engler, E. Ganssauge, M. Schweizer: DESY Report 67/34 (1967)
108 W. Bartel, B. Dudelzak, H. Krehbiel, J. Mc Elroy, U. Meyer Berkhout, W. Schmidt, V. Walther, G. Weber: Phys. Lett. 27B, 660 (1968)
109 M. Köbberling, J. Moritz, K.H. Schmidt, D. Wegener, D. Zeller, J. Bleckwenn, F.H. Hermilich: Nucl. Phys. B82, 201 (1974)
110 S. Stein, W.B. Atwood, E.D. Bloom, R.L.A. Cottrel, H. de Staebler, C.L. Jordan, H.G. Piel, C.Y. Prescott, R. Siemann, R.E. Taylor: Phys. Rev. D12, 1884 (1975)
111 T.A. Armstrong, W.R. Hogg, G.M. Lewis, A.W. Robertson, G.R. Brookers, A.S. Clough, J.H. Freeland, W. Galbraith, A.F. King, W.R. Rawlinson, N.R.S. Tait, J.C. Thompson, D.W.L. Tolfree: Phys. Lett. 34B, 535 (1971)
112 F.W. Brasse, W. Flanger, J. Gayler, S.P. Goel, R. Haidon, M. Merkwitz, H. Wriedt: Nucl. Phys. B110, 413 (1976)
113 J.C. Alder, F.W. Brasse, W. Fehrenbach, J. Gayler, S.P. Goel, R. Haidan, V. Korbel, J. May, M. Merkwitz, A. Nurimba: Nucl. Phys. B105, 253 (1976)
114 J. Gayler: Thesis, DESY F21-71/2 (1971)
J. May: Thesis, DESY F21-71/3 (1971)
115 W. Bartel, B. Dudelzak, H. Krehbiel, J. Mc Elroy, U. Meyer Berkhout, W. Schmidt, V. Walther, G. Weber: Phys. Lett. 28B, 573 (1971)
116 W. Bartel, F.W. Büsser, W.R. Dix. R. Felst, D. Harms, H. Krehbiel, P.E. Kuhlmann, J. Mc Elroy, J. Meyer, G. Weber: Phys. Lett 35B, 181 (1971)
117 K. Bätzner, U. Beck, K.H. Beckse, Ch. Berger, J. Drees, G. Knop, M. Leenen, K. Moser, Ch. Nietzel, E. Schlösser, H.E. Stier: Phys. Lett 39B, 575 (1972)

118 F.W. Brasse, W. Fehrenbach, W. Flauger, K.H. Frank, J. Gayler, V. Korbel, J. May, P.D. Zimmermann, F. Ganssauge: *Int. Symp. on Electron and Photon Interactions at High Energy*, Ithaca, N.Y., 23-27 August (1971)
119 J.C. Alder, F.W. Brasse, E. Chazelas, W. Fehrenbach, W. Flauger, K.H. Frank, E. Ganssauge, J. Gayler, W. Krechlok, V. Korbel, J. May, M. Merkwitz, P.D. Zimmermann: Nucl. Phys. B48, 487 (1972)
120 W. Albrecht, F.W. Brasse, H. Dorner, W. Flauger, K.H. Frank, J. Gayler, H. Hetschin, V. Korbel, J. May, E. Ganssauge: DESY 69/46, November (1969)
121 B. De Tollis, F. Nicolo: Nuovo Cimento 48A, 281 (1967)
122 Yu.S. Surovtsev, G.F. Tkebuchava: Preprint, Joint Institute Nucl. Res., Dubna, E2-8018 (1974)
123 W.R. Frazer: Phys. Rev. 115, 1763 (1959)
124 G.F. Chew, F.E. Low: Phys. Rev. 113, 1640 (1959)
125 R.C.E. Devenish, D.H. Lyth: Phys. Rev. D5, 47 (1972); D6, 2067 (1972)
126 B.H. Kellett, C. Verzegnassi: Nuovo Cimento 20A, 194 (1974)
127 E. Amaldi, B. Borgia, P. Pistilli, M. Balla, G.V. Di Giorgio, A. Giazotto, S. Serbassi, G. Stoppini: Nuovo Cimento 65A, 377 (1970)
128 D.R. Botterill, D.W. Braben, R. Kikuchi, P.R. Norton, A. Del Guerra, A. Giazotto, M.A. Giorgi, A. Stefanini: Phys. Lett. 45B, 405 (1973)
129 A. Del Guerra, A. Giazotto, M.A. Giorgi, A. Stefanini, B.H. Kellett, C. Verzegnassi: Phys. Lett. 50B, 487 (1974)
130 F.A. Berends: Phys. Rev. D1, 2590 (1970)
131 C.N. Brown, C.R. Canizares, W.E. Cooper, A.M. Eisner, G.J. Feldmann, C.A. Lichtenstein, L. Litt, W. Lookeretz, V.B. Montana, F.M. Pipkin: Phys. Rev. D8, 92 (1973)
132 C.J. Bebeck, C.N. Brown, M. Herzlinger, S. Holmes, C.A. Lichtenstein, F.M. Pipkin, L.K. Sisterson, D. Andrews, K. Berkelmann, D.G. Cassel, D.L. Hartill: Phys. Rev. 9D, 1229 (1974)
133 Other papers reporting on coincidence electroproduction experiments are:
C. Driver, K. Heinloth, K. Höhne, G. Hofmann, P. Karow, D. Schmidt, G. Specht, J. Rathje: Phys. Lett. 35B, 77 (1971)
A.B. Clegg, F. Foster, G. Hughes, R. Siddle, J. Allison, B. Dickinson, E. Evangelidis, M. Ibbotson, R. Lawson, P.S. Kummer, R.S. Meaburn, H.E. Montgomery, W.J. Shuttleworth, A. Sofair: Lettere al Nuovo Cimento 1, 1026 (1962); Nucl. Phys. B42, 369 (1972)
134 A. Del Guerra, A. Giazotto, M.A. Giorgi, A. Stefanini, D.R. Botterill, D.W. Braben, P.R. Norton: Phys. Lett. 45B, 409 (1973)
135 A.S. Esaulov, A.S. Omelaenko, A.M. Pilipenko, Yu.I. Titov: paper n. 504 presented at the *XVIII Int. Conf. on High Energy Physics*, Tbilisi, 15-21 July (1976). This paper contains a complet list of references to the work made previously at the Kharkov electron linear accelerator. Although rather accurate, they are all single-arm experiments. See also Rev. 158.
136 P. Brauel, F.W. Büsser, Th. Canzler, D. Cords, W.R. Dix, R. Felst, G. Grindhammer, W.D. Kollmann, H. Krehbiel, J. Meyer, G. Weber: Phys. Lett. 50B, 507 (1974)
137 D.R. Botterill, A. Del Guerra, A. Giazotto, M.A. Giorgi, G. Matone, H.E. Montgomery, P.R. Norton, A. Stefanini: Nucl. Phys. B116, 65 (1977)
138 R.C.E. Devenish, D.H. Lyth: Nucl. Phys. B93, 109 (1975)
139 J.C. Alder, F.W. Brasse, E. Chazelas, W. Fehrenbach, W. Flauger, K.H. Frank, E. Ganssauge, J. Gayler, V. Korbel, J. May, M. Merkwitz, A. Courau, G. Tristram, J. Valentin: Nucl. Phys. B46, 573 (1972)
140 J.P. Perez-y-Jorba, P. Bounin, J. Chollet: Phys. Lett. 11, 350 (1964)
W.W. Ash, K. Berkelmann, C.A. Lichtenstein, A. Ramanauskas, R.H. Siemann: Phys. Lett. 24B, 165 (1965)
C.W. Akerlof, W.W. Ash, K. Berkelman, M. Tigner: Phys. Rev. Lett. 14, 1036 (1965)
R. Kikuchi, K. Baba, S. Kaneko, H. Huke, Y. Kobayashi, T. Yamakawa: Nuovo Cimento 43A, 1178 (1966)
K. Baba, N. Kajiura, S. Kaneko, K. Huke, R. Kikuchi, Y. Kobayashi, T. Yamakawa: Nuovo Cimento 59A, 53 (1969)
N. Kajiura, K. Baba, R. Hamatsu, N. Ishihara, S. Kaneko, T. Katsura, T. Ohsugi, S. Fukui, M. Hongoh, T. Ohsuka, K. Ueno, T. Katayama, Y. Kobayashi: Nuovo Cimento Lett. 3, 714 (1970)

F.W. Brasse, E. Chazelas, W. Fehrenbach, K.H. Frank, J. Gayler, V. Korbel, J. May, A. Courau, G. Tristram, E. Ganssauge: Contribution to the 1971 International Symposium on Electron and Photon Interactions at High Energies, (Cornell, 1971)
S. Galster, G. Hartwig, H. Klein, J. Moritz, K.H. Schmidt, W. Schmidt-Parzefall: Phys. Rev. D5, 519 (1972)
R.D. Hellings, D. Wegener, J. Allison, A.B. Clegg, F. Foster, G. Hughes, P. Kummer, R. Siddle, B. Dickinson, M. Ibbotson, R. Lawson, H.E. Montgomery, W.J. Shuttleworth, A. Sofair, J. Fannon: Daresbury Report, DNL/P 65
W. Bartel, B. Dudelzak, H. Krehbiel, J. Mc Elroy, U. Meyer-Berkhout, W. Schmidt, V. Walther, G. Weber: Phys. Lett. 28B, 148 (1968)
141 W. Albrecht, F.W. Brasse, H. Dorner, W. Fehrenbach, W. Flauger, K.H. Frank, J. Gayler, V. Korbel, J. May, P.D. Zimmermann, A. Courau, A. Diaczek, J.C. Dumas, G. Tristram, J. Valentin, C. Aubret, E. Chazelas, E. Ganssauge: Nucl. Phys. B25, 1 (1971); B27, 615 (1971)
142 J.C. Adler, F.W. Brasse, E. Chazelas, W. Fehrenbach, W. Flauger, K.H. Frank, E. Ganssauge, J. Gayler, V. Korbel, J. May, M. Kerkwitz, C. Courau, G. Tristram, J. Valentin: Nucl. Phys. B46, 573 (1972)
143 R. Siddle, B. Dickinson, M. Ibbotson, R. Lawson, H.E. Montgomery, V.P.R. Nuthakki, O.T. Tumer, W.J. Shuttleworth, A. Sofair, R.D. Hellings, J. Allison, A.B. Clegg, F. Foster, G. Hughes, P.S. Kummer, J. Fannon: Nucl. Phys. B35, 93 (1971)
144 F. Gutbrod: DESY 69/22 (1969); 69/33 (1969)
145 G. Fischer, H. Fischer, G.V. Holtey, H. Kämpgen, G. Knop, P. Schulz, H. Wessels, W. Braunschweig, H. Genzel, R. Wedemeyer: Nucl. Phys. B16, 93 (1970)
146 P. Noelle, W. Pfeil, D. Schwela: Nucl. Phys. B26, 461 (1971)
147 R.L. Crawford: Nucl. Phys. B28, 573 (1971)
148 R.C.E. Devenish, D.H. Lyth: DNPL/P 98 (1971)
149 W.W. Ash, K. Berkelman, C.A. Lichtenstein, A. Ramanauskas, R.H. Siemann: Phys. Lett. 24B, 165 (1967)
150 E. Amaldi, M. Beneventano, B. Borgia, F. De Notaristefani, A. Frondaroli, P. Pistilli, I. Sestili, M. Severi: Phys. Lett. 41B, 216 (1972)
151 D.R. Botterill, D.W. Braben, P.R. Norton, A. Del Guerra, A. Giazotto, M.A. Giorgi, A. Stefanini: Lettere al Nuovo Cimento 10, 629 (1972)
A. Del Guerra, A. Giazotto, M.A. Giorgi, A. Stefanini, D.R. Botterill, D.W. Braben, D. Clarke, P.R. Norton: Nucl. Phys. B99, 253 (1975)
152 A. Del Guerra, A. Giazotto, M.A. Giorgi, A. Stefanini, D.R. Botterill, H.E. Montgomery, P.N. Norton, G. Matone: Nucl. Phys. B107, 65 (1976)
153 P. Spillantini, V. Valente: CERN Report: CERN-Hera 70-1 (1970)
154 B.J. Read: Nucl. Phys. B74, 482 (1974)
155 M.I. Adamovich, V.G. Laranova, A.I. Lebedev, S.P. Kharlamov, F.R. Yagudina: J. Nucl. Phys. (U.S.S.R.) 2, 135 (1965); Soviet J. Nucl. Phys. 2, 95 (1966)
156 G. Bardin, J. Duclos, J. Julien, A. Magnon, B. Michel, J.C. Montret: Lettere al Nuovo Cimento 13, 485 (1975); Compete Rendu d'Activité (1 Sept. 1953 - 30 Sept. 1976), Département de Physique Nucléaire, CEN de Saclay.
G. Bardin, J. Duclos, A. Magnon, B. Michel, J.C. Montret: Nucl. Phys. B120, 45 (1977)
157 Beneventano, B. Borgia, A. Capone, F. De Notaristefani, M. De Vincenzi, E. Longo, M. Mattioli, P. Pistilli, I. Sestili, M. Severi, G. Picozza: Phys. Lett. 62B, 114 (1976)
158 A.S. Saulov, A.M. Pilipenko, Yu.I. Titov: Nucl. Phys. B136, 511 (1978)
159 N. Paver, P. Pistilli, I. Sestili, C. Verzegnassi: Nuovo Cimento 34A, 182 (1976)
160 G. Furlan, N. Paver, C. Verzegnassi: to be published on IL NUOVO CIMENTO
161 N.P. Samios: Phys. Rev. 121, 272 (1961)
162 H. Kolrak: Nuovo Cimento 20, 115 (1961)
163 T.D. Blokhintseva, V.A. Grebinnik, V.A.Z. Zhukov, A.V. Krautson, G. Libman, L.L. Nemenov, G.I. Selivanov, Yuan Jung-Fang: Yadernaya Fizika 3, 779 (1966)
164 D.A. Geffen: Phys. Rev. 125, 1745 (1962)
J. Loubaton, J. Tran Thanh Van: Nucl. Phys. 2B, 342 (1967)

165 Yu. K. Akimov, L.S. Vertogradov, A.V. Dem'Yanov, A.V. Kuptsov, L.L. Nemenov, D.M. Khazins, Yu.M. Chirkin, Yu.D. Prokoshkin, N.M. Agababyan, I.A. Keropyan, G.G. Mkrtchyan, S.F. Berezhnev, A.V. Kulikov, G.I. Smirnov: Soviet J. Nucl. Phys. 13, 425 (1971)

166 R. Garland: Thesis; Department of Physics of the Columbia University, R-764; CU-294; News 188 (1971)

167 Yu.S. Surovtsev, F.G. Tkebuchava: JINR, Dubna, preprint R2-4561 (1969)

168 S.G. Petrova: JINR, Dubna, preprint R2-7037 (1973)

169 G.I. Smirnov, N.M. Sciumejko: JINR, Dubna, preprint R2-6871 (1973)

170 Yu.S. Surotsev, F.G. Tkebuchava: Soviet J. Nucl. Phys. 16, 713 (1973)

171 Yu. K. Kulish: Soviet J. Nucl. Phys. 16, 605 (1973)

172 A.M. Baldin, V.A. Suleimanov: JINR, Dubna, preprint R2-7096 (1973)

173 A. Bietti, S. Petrarca: Nuovo Cimento 22A, 595 (1974); Lettere al Nuovo Cimento 13, 539 (1975)

174 G. Furlan, N. Paver, C. Verzegnassi: Nuovo Cimento A32, 75 (1976)

175 S.F. Berezhnev, L.S. Vertogradov, A.V. Dem'Yanov, A.V. Kulikov, A.V. Kuptsov, G.G. Mkrtchyan, L.L. Nemenov, G.I. Smirnov, D.A. Khazins, Yu.M. Chirkin: Soviet J. Nucl. Phys. 16, 99 (1973)
S.F. Berezhnev, A.V. Dem'Yanov, A.V. Kulikov, A.V. Kuptsov, V.P. Kurochkin, G.G. Mkrtchyan, L.L. Nemenov, Zh.P. Pustyl'nik, G.I. Smirnov, A.G. Fedunov, D.M. Khazins: Soviet J. Nucl. Phys. 17, 44 (1973); JINR, Dubna, preprint P1-6624 (1972); JINR, Dubna, preprint P1-6934 (1973); Soviet J. Nucl. Phys. 18, 53 (1974). For a review see the Rapporteur talk by G. Wolf at the 1975 Symposium on Lepton and Photon Interactions at High Energies (Stanford, August 1975) p. 795

176 V.V. Alizade, S.F. Berezhnev, A.V. Dem'Yanov, A.V. Kuptsov, V.P. Kurochkin, L.L. Nemenov, Zh.P. Pustyl'nik, G.I. Smirnov, D.M. Khazins: JINR, Dubna, preprint P1-9478 (1976)

177 G.J. Gounaris, J.J. Sakurai: Phys. Rev. Lett. 21, 244 (1968)

178 R.E. Cutkosky, F. Zachariasen: Phys. Rev. 103, 1108 (1956)

179 P. Stichel, M. Scholz: Nuovo Cimento 34, 1381 (1964)

180 For a review see D. Luke, P. Söding: Springer Tracts in Modern Physics 59, 39 (1971)

181 F.A. Berends, R. Gastmans: Phys. Rev. D5, 204 (1972)
P. Levi, W. Schmidt: in *Int. Symp. on Lepton and Photon·Interaction at High Energy* (Stanford, August 1975)

182 A. Bartl, W. Majerotto, D. Schildknecht: Nuovo Cimento 12A, 703 (1972)

183 C. Driver, K. Heinloth, K. Höhne, G. Hofmann, P. Karow, D. Schmidt, G. Spocht: Nucl. Phys. B32, 45 (1971)

184 P. Carruthers, K.W. Huang: Phys. Lett. 24B, 464 (1967)
See also T. Ebata: Phys. REv. 154, 1341 (1967)
P. Narayanaswamy, B. Renner: Nuovo Cimento 53A, 107 (1968)

185 S.L. Adler, W. Weisberger: Phys. Rev. 169, 1392 (1968)

186 A. Bartl, N. Paver, C. Verzegnassi, R. Wittmann: Nuovo Cimento, 45A, 457 (1978)
See also A. Bartl, N. Paver, S. Petrarca, C. Verzegnassi: "Lettere al Nuovo Cimento" 18, 588 (1977)

187 I. Dammann, C. Driver, K. Heinloth, G. Hofmann, F. Janata, P. Karow, D. Lüke, D. Schmidt, G.S. Specht: Nucl. Phys. B54, 355 (1973)

188 P. Joos, A. Ladage, H. Meyer, D. Notz, P. Stein, G.Wolf, S. Yellin, C. Benz, G. Drews, D. Hoffmann, J. Knobloch, W. Kraus, H. Nagel, E. Rabe, C. Sander, W.D. Schlatter, H. Spitzer, K. Wacker, P. Winkler, I.J. Bloodworth, C.K. Chen, J. Knowles, D. Martin, J.M. Scarr, I.O. Skillicorn, K. Smith: Phys. Lett. 52B, 481 (1974). See also /24/

189 H.E. Montgomery, J. Allison, B. Dickinson, E. Evangelides, M. Ibbotson, R.S. Lawson, R.S. Meaburn, W.J. Shuttleworth, A. Sofair, F. Foster, G. Hughes, P.S. Kummer, R. Siddler: Nucl. Phys. B51, 377 (1973)

190 P. Joos, A. Ladage, H. Meyer, P. Söding, P. Stein, G. Wolf, S. Yellin, C.K. Chen, J. Knowles, D. Martin, J.M. Scarr, I.O. Skillicorn, K. Smith, C. Benz, G. Drews, D. Hoffmann, J. Knobloch, W. Kraus, H. Nagel. E. Rabe, C. Sander, W.D. Schlatter, H. Spitzer, W. Wacker: Phys. Lett. 62B, 230 (1976)
191 P.S. Kummer, E. Ashburner, F. Foster, G. Hughes, R. Siddle, J. Allison, B. Dickinson, E. Evangelides, M. Ibbotson, R.S. Lawson, R.S. Meaburn, H.E. Montgomery, W.J. Shuttleworth: Phys. Rev. Lett. 30, 873 (1973)
192 U. Beck, K.H. Becks, V. Burkert, J. Dreus, B. Dreshbachn, B. Gerhardt, G. Knop, H. Kolanoski, M. Leenen, K. Moser, H. Müller, Ch. Nietzel, J. Päsler, K. Rith, M Rosenberg, V. Sauerwein, E. Schlösser, H.E. Stier: Phys. Lett. 51B, 103 (1974)
193 J.C. Adler, F.W. Brasse, W. Fehrenbach, J. Gayler, R. Haidan, G. Glöe, S.P. Goel, V. Korbel, W. Krechlock, J. May, M. Merkwitz, R. Schmitz, W. Wagner: Nucl. Phys. B91, 386 (1975). The separation of σ_L and σ_T in ψ-electroproduction at the resonance S_{11} (1535) has been recently performed by F.W. Brasse et al. (DESY report 77/73, November 1977).
194 R.P. Bajpai, A. Donnachie: Nucl. Phys. B12, 274 (1969)
195 F. Ravndal: Phys. Rev. D4, 1466 (1971)
196 R.G. Lipes, Phys. Rev. D5, 2849 (1972)
197 C.N. Brown, C.R. Canizares, W.E. Cooper, A.M. Eisner, G.J. Feldman, C.A. Lichtenstein, L. Litt, W. Lockeretz, V.B. Montana, F.M. Pipkin, N. Hicks: Phys. Rev. Lett. 28, 1086 (1972)
198 T. Azemoon, I. Dammann, C. Driver, D. Lüke, G. Specht, K. Heinloth, H. Ackermann, E. Ganssauge, F. Janata, D. Schmitz: Nucl. Phys. B95, 77 (1975)
199 C.J. Bebek, C.N. Brown, M. Herzlinger, S. Holmes, C.A. Lichtenstein, F.M. Pipkin, D. Andrews, K. Berkelman, D.G. Cassel, D.L. Hartill, N. Hicks: Phys. Rev. Lett. 32, 21 (1974)
200 A.M. Boyarski, F. Bulos, W. Busza, R. Diebold, S.D. Ecklund, G.E. Fisher, Y. Murata, J.R. Rees, B. Richter, W.S.C. Williams: Phys. Rev. Lett. 22, 1131 (1969)
201 See, for example, G. Ebel, D. Julius, A. Müllensiefen, H. Pilkuhn, W. Schmidt, F. Steiner, G. Kramer, G. Schierholz, B.R. Martin, J. Pisut, M. Roos, G. Oades, J.J. De Swart: Springer Tracts in Modern Physics 55, 239 (1970)
202 A. Bartl, W. Majerotto: Nucl. Phys. B90, 285 (1975)
203 D. Basu, R.N. Chaudhuri: Phys. Rev. 175, 2075 (1968)
204 F. von Hippel, J.K. Kim: Phys. Rev. D1, 151 (1970)
H. Fritzsch, M. Gell-Mann, H. Leutwyler: unpublished
205 R.L. Walker: Phys. Rev. 182, 1729 (1969)
206 A. Bartl, W. Majerotto: Nucl. Phys. B62, 267 (1973)
207 A. Actor: Ann. Phys. (N.Y.) 84, 318 (1974)
208 N. Dombey: Rev. Mod. Phys. 41, 236 (1969)
209 N. Christ, T.D. Lee: Phys. Rev. 143, 1310 (1966)
210 J.R. Chen, J. Sanderson, J.A. Appel, G. Gladding, M. Goitein, K. Hanson, D.C. Imrie, T. Kirk, R. Madaras, R.V. Pound, L. Price, R. Wilson, C. Zajde: Phys. rev. Lett. 21, 1279 (1968)
211 S. Rock, M. Borghini, O. Chamberlain, R.Z. Fuzesy, C.C. Morehouse, T. Powell, G. Shapiro, H. Weisberg, R.L.A. Cottrell, J. Litt, L.W. Mo, R.E. Taylor: Phys. Rev. Lett. 24, 748 (1970)
212 F.A. Berends: Phys. Rev. D1, 2590 (1971)
213 See, for instance, G. Zweig: Nuovo Cimento 32, 689 (1964) (photoproduction only)
M. Boiti, F. Pempinelli: Nuovo Cimento 54A, 108 (1968)
J.S. Ball, M. Jacob: Nuovo Cimento 54A, 620 (1968)
214 See "Review of Particle Properties", Particle Data Group, Review of Modern Physics 48, Part. II (1976)
215 See, for instance, Yu.V. Novozhilov: Introduction to Elementary Particle Physics (Pergamon Press, 1975) Chap. 13
G. Giacomelli: Physics Reports 23C, 123 (1976)
216 F. Nicolo, G.C. Rossi: Nuovo Cimento 56A, 207 (1968)

217 The first example can be found in: S. Fubini, Y. Nambu, V. Wataghin: Phys. Rev. 111, 329 (1958)
See also F.A. Berends, G.B. West: Phys. Rev. 188, 2538 (1969)
218 S.F. Berezhnev, T.D. Blokhintseva, V.A. Demyanov, A.V. Kuptsev, V.P. Kurochkin, L.L. Nemenov, S.I. Smirnov, D.M. Khazius: Soviet J. Nucl. Phys. 24, 591 (1976); 26, 547 (1977)
219 F.G. Tkebuchava: JINR, Dubna, preprint P2-11152 (1978), to be published on IL NUOVO CIMENTO

Springer Tracts in Modern Physics

Editor: G. Höhler
Associate Editor: E. A. Niekisch

Volume 62

Photon-Hadron Interactions I

International Summer Institute in Theoretical Physics, DESY, July 12–24, 1971

1972. 46 figures. VII, 147 pages
ISBN 3-540-05757-9

Contents:
K. Huang: Duality and the Pion Electromagnetic Form Factor. – *K. Huang:* Deep Inelastic Hadronic Scattering in Dual-Resonance Model. – *F. M. Renard:* p–a Mixing. – *V. Rittenberg:* Scaling in Deep Inelastic Scattering with Fixed Final States. – *H. R. Rubinstein:* Duality for Real and Virtual Photons. – *P. V. Landshoff:* Duality in Deep Inelastic Electroproduction. – *K. Schilling:* Some Aspects of Vector Meson Photoproduction on Protons. – *A. Donnachie:* Exotic Electromagnetic Currents. – *R. Jackiw:* Canonical Light-Cone Commutators and Their Applications.

Volume 63

Photon-Hadron Interactions II

International Summer Institute in Theoretical Physics, DESY, July 12–24, 1971

1972. 97 figures. V, 189 pages
ISBN 3-540-05813-3

Contents:
J. Frøyland: High Energy Photoproduction of Pseudoscalar Mesons. – *K. Schilling:* Some Aspects of Vector Meson Photoproduction on Protons. – *D. Schildknecht:* Vector Meson Dominance, Photo- and Electroproduction from Nucleons. – *F. M. Renard:* p–ω Mixing. – *A. Donnachie,* Exotic Electromagnetic Currents. – *A. P. Contogouris:* Regge Analysis and Dual Absorptive Model. – *P. D. B. Collins, F. D. Gault:* The Eikonal Model for Regge Cuts in Pion-Nucleon Scattering.

Volume 67
S. Ferrara, R. Gatto, A. F. Grillo

Conformal Algebra in Space-Time and Operator Product Expansion

1973. IV, 69 pages
ISBN 3-540-06216-5

Contents:
Introduction to the Conformal Group in Space-Time. – Broken Conformal Symmetry. – Restrictions from Conformal Covariance on Equal-Time Commutators. – Manifestly Conformal Covariant Structure of Space-Time. – Conformal Invariant Vacuum Expectation Values. – Operator Products and Conformal Invariance on the Light-Cone. – Consequences of Exact Conformal Symmetry on Operator Product Expansions. – Conclusions and Outlook.

Volume 79

Elementary Particle Physics

1976. 37 figures. VI, 145 pages
ISBN 3-540-07778-2

Contents:
H. Rollnik, P. Stichel: Compton Scattering: Compton Scattering in the Resonance Region. High-Energy Compton Scattering. – *E. Paul:* Status of Inference Experiments with Neutral Kaons: Interference Effects in a Beam of Coherent K_S^0 and K_L^0 and Possibilities of Measuring Them. K_S^0 Lifetime. $(K_L^0 - K_S^0)$ Mass Difference. Measurement of CP Violation in the Two-Pion Decay Modes. Search for CP Violation in Three-Body Decay Modes. Test of the $\Delta S = \Delta Q$ Rule in the Semileptonic Decay Modes. Analysis of the CP Violation Data Considering Unitarity. Possibilites of Explaining CP Violation.

Springer-Verlag
Berlin
Heidelberg
New York

Topics in Current Physics

Founded by Helmut K. V. Lotsch

Volume 1

Beam-Foil Spectroscopy

Editor: S. Bashkin

1976. 91 figures. XIII, 318 pages
ISBN 3-540-07914-9

Contents:
S. Bashkin: Introduction. – *S. Bashkin:* Experimental Methods. – *I. Martinson:* Studies of Atomic Spectra by the Beam-Foil Method. – *L. J. Curtis:* Lifetime Measurements. – *O. Sinanoğlu:* Theoretical Oscillator Strengths of Neutral, Singly-Ionized, and Multiply-Ionized Atoms. – *W. Wiese:* Regularities of Atomic Oscillator Strenghts in Isoelectronic Sequences. – *W. Whaling:* Applications to Astrophysics: Absorption Spectra. – *L. J. Heroux:* Applications of Beam-Foil Spectroscopy to the Solar Ultraviolet Emission Spectrum. – *R. Marrus:* Studies of Hydrogen-Like and Helium-Like Ions of High Z. – *J. Macek:* Coherence Alignment and Orientation. – *D. Burns:* Phenomena in the Beam-Foil Light Source. – *I. A. Sellin:* The Measurement of Autobionizing Ion Levels and Lifetimes by tast Projectile Electron Spectroscopy.

Volume 2

Modern Three-Hadron Physics

Editor: A. W. Thomas

1977. 30 figures. XI, 250 pages
ISBN 3-540-07950-5

Contents:
I. R. Afnan, A. W. Thomas: Fundamentals of Three-Body Scattering Theory. – *L. R. Dodd:* Analytic Structure of On-Shell Three-Body Amplitudes. – *R. D. Amado:* Theory of Three-Body Final States. – *D. D. Brayshaw:* The Boundary Condition Method. – *R. Aaron:* A Relativistic Three-Body Theory. – *E. F. Redish:* Applications of Three-Body Methods to Many-Body Hadronic Systems.

Volume 5

Structure and Collisions of Ions and Atoms

Editor: I. A. Sellin

1978. 157 figures, 17 tables. XI, 350 pages
ISBN 3-540-08576-9

Contents:
I. A. Sellin: Introduction. – *S. Brodsky, P. J. Mohr:* Quantum Electrodynamics in Strong and Supercritical Fields. – *L. Armstron jr.:* Relativistic Effects in Highly Ionized Atoms. – *J. S. Briggs, K. Taulbjerg:* Theory of Inelastic Atom-Atom Collisions. – *N. Stolterfoht:* Excitation in Energetic Ion-Atom Collisions Accompanied by Electron Emission. – *P. Mokler, F. Folkmann:* X-Ray Production in Heavy Ion-Atom Collisions. – *I. A. Sellin:* Extensions of Beam Foil Spectroscopy. – *S. Datz:* Atomic Collisions in Solids.

Springer-Verlag
Berlin
Heidelberg
New York